52 Advances in Biochemical Engineering Biotechnology

Managing Editor: A. Fiechter

W0246105

Springer-Verlag Berlin Heidelberg GmbH

Microbial and Enzymatic Bioproducts

With Contributions by
M. Hiroto, H. Nishimura, Y. Kodera,
F. Kawai, P. Rusin, H. L. Ehrlich,
S. Y. Lee, H. N. Chang, S. A. Markov,
M. J. Bazin, D. O. Hall, A. L. Gutman,
M. Shapira, Y. Inada, A. Matsushima

With 85 Figures and 36 Tables

 Springer

ISBN 978-3-662-14875-4   ISBN 978-3-540-49190-3 (eBook)
DOI 10.1007/978-3-540-49190-3

Library of Congress Catalog Card Number 72-152360

Springer-Verlag Berlin Heidelberg 1995
Originally published by Springer-Verlag Berlin Heidelberg New York in 1995
Softcover reprint of the hardcover 1st edition 1995

Typesetting: Macmillan India Ltd., Bangalore-25
SPIN: 10122579   02/3020 - 5 4 3 2 1 0 - Printed on acid-free paper

Attention all "Enzyme Handbook" Users:

A file with the complete volume indexes Vols. 1 through 10 in delimited ASCII format is available for downloading at no charge from the Springer EARN mailbox. Delimited ASCII format can be imported into most databanks.

The file has been compressed using the popular shareware program "PKZIP" (Trademark of PKware INc., PKZIP is available from most BBS and shareware distributors).

This file distributed without any expressed or implied warranty.

To receive this file send an e-mail message to:
SVSERV@DHDSPRI6.BITNET.
The message must be: "GET/ENZHB/ENZ_HB.ZIP".

SPSERV is an automatic data distribution system. It responds to your message. The following commands are available:

HELP	returns a detailed instruction set for the use of SVSERV,
DIR (*name*)	returns a list of files available in the directory "name",
INDEX (*name*)	same as "DIR"
CD <*name*>	changes to directory "name",
SEND <*filename*>	invokes a message with the file "filename"
GET <*filename*>	same as "SEND".

Table of Contents

Developments in Microbial Leaching –
Mechanisms of Manganese Solubilization

Patricia Rusin[1] and Henry Ehrlich[2]
[1] Department of Soil and Water Science, University of Arizona, Tucson, Arizona, 85721, USA
[2] Department of Biology, Rensselaer Polytechnic Institute, Troy, New York 12180, USA

Microorganisms can be used to solubilize manganese in manganiferous oxide ores. The organisms usually use manganese as a terminal electron acceptor reducing it to Mn(II). Manganese-reducing organisms have been isolated from freshwater and ocean sediments, marine nodules, and ore samples. These organisms require an organic source of carbon which can be provided in the form of molasses or other food industry waste products. The manganese-reducing bacteria and fungi can be used to extract manganese from low-grade manganiferous ore, to separate manganese from iron in ferromanganiferous ore, and to release silver from refractory manganese oxide ores.

Advances in Biochemical Engineering/
Biotechnology, Vol. 52
Managing Editor: A. Fiechter
© Springer-Verlag Berlin Heidelberg 1995

1 Introduction

Most authors who have described microbial manganese reduction have done so from an ecological point of view. That is, they have been interested in reduction as part of the global manganese biogeochemical cycle. However, some authors have also studied microbial manganese reduction as a tool for metal extraction.

World reserves in commercial manganiferous deposits has been estimated at 10^9 metric tons of manganese [1]. The most important ores are oxides. Most of these minerals contain Mn(IV) in forms such as pyrolusite, psilomelane, and cryptomelane. Other minerals such as manganite, brownite, and hausmannite contain forms of Mn(III). The Mn(IV) and Mn(III) mineral forms found in these ores are insoluble and manganese must be reduced to Mn(II) to achieve extraction. The most important terrestrial deposits in descending order of size are found in South Africa, the former Soviet Union, Gabon, Australia, Brazil, India, China, Ghana, Mexico, and Morocco. In many cases the rich deposits with manganese concentrations in excess of 35% are not found in the industrialized countries. In addition to these reserves, ferromanganese nodules on the ocean floor are also a potential source of manganese. Reserves of ferromanganese concretions and crusts found on the ocean floors have been estimated at approximately 10^9 tons [2].

Numerous manganiferous deposits occur throughout the world which remain undeveloped due to their refractory nature. In some cases, silver is entrapped within minerals such as pyrolusite (MnO_2) and is not amenable to cyanide extraction [3]. Bacteria that can solubilize these oxide minerals can be used to free entrapped metals.

Manganese is used in ferromanganese and other chemical industries and as a strategically important metal in the manufacture of steel. It can also be used in agricultural fertilizers and as battery grade manganese oxides.

Microorganisms (bacteria and fungi) that can reduce particulate Mn(IV) and Mn(III) to soluble Mn(II) have been isolated from such diverse habitats as marine ferromanganese nodules [4, 5], ocean sediments [6], freshwater sediments [7–9], and manganiferous ores [10–12]. Many of the MnO_2-reducing bacteria have been shown to attack solid material directly [8, 13, 14]; thus they are able to reduce manganese and solubilize minerals such as pyrolusite and birnessite.

2 Isolation of Manganese-Reducing Microorganisms

Manganese-solubilizing bacteria can be isolated from a variety of microbial habitats including active or abandoned mine sites. For example, manganese-reducing bacteria can be isolated from pregnant and barren pond sediments and water, inlets to ponds, and ore samples.

Enrichment of field samples can be carried out in a growth medium amended with $10 \ mmol \ l^{-1}$ glucose using the soft agar overlay methods of Myers and Nealson [15]. Alternatively, field samples can be directly diluted into a semi-solid growth medium as described by Rusin et al. [13]. In most cases, cultures should be incubated at 25–35 °C for up to 30 days. Microbial reduction of MnO_2 can be detected by a color change from black to clear. It is often difficult to detect colony formation using the overlay method. However, individual colonies are readily apparent using the semi-solid growth medium technique. Subsamples of the cultures in which manganese reduction is detected can be transferred to Tryptic Soy Agar (TSA, Difco) for bacterial isolation. Purified bacterial isolates can be tested for the ability to reduce Mn(IV) by growing them on TSA for 48 h, suspending colonies in sterile saline (0.75%) to a McFarland 0.5 standard [16] turbidity equivalent (approx. 1×10^8 colony forming units (cfu) ml^{-1}), and screening by the semi-solid medium technique for manganese reduction. Those isolates which reduce manganese can be maintained on TSA and lyophilized for further study.

Environmental samples for microbial analyses are usually transported to the laboratory on ice [17]. It is important that the sample is not frozen or subjected to intense heat. Ehrlich [4] isolated manganese-reducing bacteria from ferromanganese nodules from the Pacific Ocean. Nodules were frozen or refrigerated during transport to the laboratory. The frozen nodules were sampled by surface scraping. Refrigerated nodules were sampled by collecting and crushing surface flakes. Much higher numbers of bacteria were recovered from the refrigerated nodules than from the frozen ones. This is not surprising as freezing is known to be injurious to bacteria. However, bacterial multiplication may occur at 4 °C, especially in samples likely to contain psychrotrophic organisms. Growth at low temperatures would be quite likely from samples obtained from areas such as the ocean floor with an average temperature of 3–5 °C. The most desirable procedure is to process samples without prior storage although this is often impractical in the field.

Manganese-solubilizing bacteria may require an induction period before Mn(IV) reduction is initiated. Some marine isolates cannot reduce manganese dioxide immediately upon contact unless an artificial electron carrier such as ferricyanide ($K_3Fe(CN)_6$) is present. The ferricyanide acts as a substitute for a missing component of the electron transport chain to Mn(IV) giving the bacterium time to form the necessary electron carrier [4, 14]. Thus, the addition of ferricyanide to the medium may accelerate the detection of Mn(IV) reduction by these marine bacteria.

Some of the most active manganese-reducing bacteria isolated from ore samples are members of the bacterial genus *Bacillus* [11]. Two isolates, MBX 1 and MBX 2 (formerly named D1), were further identified as *Bacillus circulans* and *Bacillus polymyxa* respectively. *Bacillus* MBX 1 exhibits requirements for growth factors including thiamine, biotin, and niacin. However, most of these bacterial strains grow well in an undefined organic medium containing molasses, whey, or PGM (Metallurgical and Biological Extraction Systems, proprietary) which provides organic carbon and growth factors. Not all manganese-solubilizing bacteria require an organic source of nitrogen. Like some other members of the genus *Bacillus* [18], isolate MBX 2 is able to incorporate atmospheric nitrogen to satisfy its metabolic nitrogen requirements. This ability is termed "nitrogen fixation" and is unique to certain members of the bacterial world. Thus, nitrogen additives are not necessary for bioleaching with MBX 2 so long as atmospheric nitrogen is available.

Bacillus MBX 1 and MBX 2 are facultative anaerobes, able to grow with or without atmospheric oxygen [11, 19]. If oxygen is available, these *Bacillus* strains will use it in a respiratory pathway as evidenced by the lethal effect of cyanide. When oxygen is absent, they will switch to anaerobic metabolism, producing ethanol as a major end-product. This bi-modal metabolic capability is termed "facultative". Being members of the genus *Bacillus*, these bacteria produce endospores (internal resting structures) which enable them to survive long periods of nutrient or moisture deprivation. If a bioreactor system is upset by fluctuations in pH, temperature, soluble metal concentrations, or other factors, the *Bacillus* population may regrow due to the presence of endospores.

Bacteria have been isolated that grow on "oxidizable" carbon sources using Mn(IV) as an electron acceptor for anaerobic respiration. One of these, strain *Shewanella putrefaciens* MR-1, was isolated from anaerobic freshwater sediments [8]. It is a facultatively anaerobic, Gram-negative rod. Isolate MR-1 is necessarily respiratory and grows anaerobically with lactate (or other electron donors), using MnO_2 as an electron acceptor. A second strain of *Shewanella* that reduces manganese via a respiratory pathway, using lactate as an electron donor, is *S. putrefaciens* sp 200 [20, 21]. This bacterium was first described as an iron reducer; further studies have shown that it also reduces manganese [11]. The marine bacterium, strain SSW_{22} has shown the ability to use acetate, succinate, and glucose as electron donors to reduce Mn(IV) under aerobic and anaerobic conditions [2].

Marine pseudomonad BIII 88 was isolated from Pacific deep sea sediment associated with a ferromanganese deposit [22]. It reduces manganese dioxide both aerobically and anaerobically at similar rates using glucose or acetate as carbon sources and electron donors. Other bacteria can also reduce manganese aerobically as well as anaerobically [9, 23].

Many other genera of bacteria have also been described that solubilize Mn(IV). A manganese-reducing bacterium designated 1-29S was isolated from a Groote Eylandt solid slime dam residue [24]. This bacterium is Gram variable, non-spore forming, motile, and forms a brown soluble pigment when grown on solid agar. Agate and Deshpande [25] leached manganese using a *Pseudomonas* sp., *Bacillus* sp., and *Arthrobacter* sp. Babenko et al. [26] leached manganese ore from the Nikopol deposit with an *Achromobacter*, while Groudev [27] has demonstrated microbial dissolution of Mn(IV) using *Alcaligenes, Bacillus, Pseudomonas* and *Aspergillus niger* (a fungus). Many other manganese-reducing bacteria are listed by Ehrlich [28].

3 Microbial Solubilization of Manganese

3.1 Direct Reduction

In most cases, the manganese-reducing bacteria described in the literature require direct physical contact to reduce manganese [8, 11, 29, 30]. There is much evidence that manganese is often reduced directly via an electron transport chain [6–8, 22, 29, 31]. Indeed, Ghiorse and Ehrlich [29] showed transmission electron micrographs showing thin sections in which the cells of *Bacillus* 29 appeared to establish an intimate contact with the Fe-Mn oxide particles. Cell fractionation experiments confirmed that the manganese reductase resided in the cell membrane. Several different mechanisms to elucidate the pathway of the electron to Mn(IV) have been proposed [32–38].

Three separate experiments were conducted showing that *Bacillus* MBX 2 also requires direct contact between the cell envelope and the ore particle for the dissimilative reduction of MnO_2 [11]. Cell-free filtrates obtained from cultures that reduced manganese did not reduce manganese when added to growth medium amended with 20 mmol l^{-1} manganese (Mn medium) even after 30 days of incubation. Furthermore, when isolate MBX 2 was inoculated into Mn medium contained in a dialysis bag (mol. wt. cutoff of 12 000) which in turn was suspended in the same medium, only manganese in the bag was reduced as evidenced by dissolution of the MnO_2. Thus reduction of manganese was not mediated by biogenic low-molecular-weight compounds such as organic acids or hydrogen peroxide. Finally, when a 0.45 μm filter was inoculated on one side with *Bacillus* MBX 2 and incubated between two layers of manganese-containing medium, manganese reduction only occurred in the medium next to the inoculated side of the filter. The 0.45 μm nominal pore diameter would allow large molecules including proteins to pass but would exclude the passage of cells. These results indicate that MBX 2 mediates manganese reduction by a cell surface associated enzyme.

In the case of *Bacillus* MBX 2, there is no evidence of energy conservation in terms of growth under anaerobic conditions. The anaerobic growth of MBX 2 was unaffected by the presence of 25 mmol l^{-1} MnO_2. This suggests that *Bacillus* MBX 2 may use Mn(IV) as an electron sink to allow for the reoxidation of nicotinamide adenine dinucleotide (NADH) rather than for electron transport (oxidative phosphorylation), similar to the mode of manganese reduction mediated by *Clostridium* sp. [30]. The primary end products from MBX 2 were acetic acid and ethanol. The concentration of acetic acid formed remained constant with or without the presence of Mn(IV). However, there was a 6-fold decrease in the quantity of ethanol produced after 3 days of incubation in the presence of Mn(IV). Isolate MBX 2 may divert reducing power from the production of ethanol to the reduction of Mn(IV) when manganese is available. However, it is also possible that the reaction system contained two populations of cells. Those not in contact with Mn(IV) may have degraded glucose (starch) forming acetic acid and ethanol. Those not in contact with Mn(IV) may have produced acetic acid through the incomplete oxidation of glucose.

Other bacteria, requiring direct contact with the mineral surface, reduce manganese aerobically [9] or oxygen is required for the adaptation of the culture to initiate MnO_2 reduction [39]. In the latter case, cells adapted to Mn(IV) reduction used MnO_2 in preference to O_2 as the terminal electron acceptor.

The marine pseudomonad strain BIII 88 reduces Mn(IV) under both aerobic and anaerobic conditions [22, 38]. In the proposed model (Fig. 1), the anaerobic or aerobic oxidation of acetate generates reducing power in the form of electrons. The electrons are fed into an electron transport chain located in the plasma membrane of the bacterial cell. The electrons are eventually transported by a carrier to the outer membrane and reduces bound Mn^{3+} to Mn^{2+}. The bound Mn^{3+} results from a disproportionation reaction at the cell surface/MnO_2 mineral surface involving Mn^{2+} in the outer cell membrane:

$$Mn^{2+} + MnO_2 + 2H_2O = 2Mn^{3+} + 4OH- . \qquad (1)$$

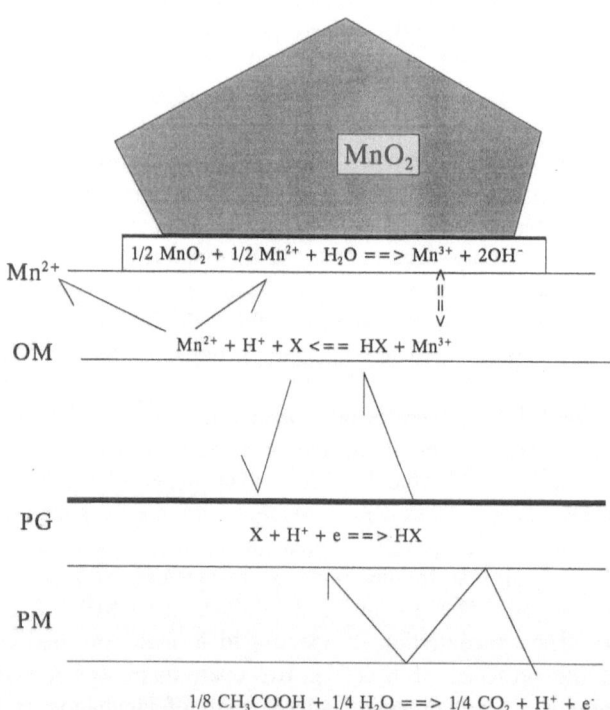

Fig. 1. Schematic representation of a model explaining the transfer of reducing power (electrons) across the interface between a bacterial cell surface and the surface of an MnO_2 particle with which the bacterium is in physical contact. OM, outer membrane; PG, peptidoglycan layer; PM, plasma membrane; X hypothetical carrier of reducing power in the cell envelope

Thus, the carrier of electrons from the electron transport system to MnO_2 is Mn^{2+}. Part of the Mn^{2+} resulting from the enzymatic reduction of Mn^{3+} diffuses into the environment and part is involved in further disproportionation of manganese dioxide. Aerobically, part of the cells reducing power is transferred to oxygen, whereas, anaerobically, all the reducing power is transferred to manganese dioxide [22].

Strains of bacteria that reduce Mn(IV) only under anaerobic conditions may use iron as an electron shuttle. Ferrous iron is much more susceptible to autoxidation at a neutral pH than manganous manganese so organisms using ferrous iron as an electron carrier could only do so under anaerobic conditions. *Shewanella putrefaciens* contains a *c*-type cytochrome in its outer membrane when grown anaerobically but not when grown aerobically [32]. This *c*-type cytochrome could represent the iron or it could be a carrier for the reduction of ferric iron in a polynuclear complex similar to that found in the outer membrane of *T. ferrooxidans* [40–42].

With iron as the electron shuttle, the standard free energy potentially available to the cell per mole of MnO_2 reduced by acetate is -53.4 kcal. With manganese as the electron shuttle the standard free energy potentially available to the cell per mole of Mn^{3+} reduced to Mn^{2+} is -33.2 kcal. Therefore, iron is theoretically a more efficient electron shuttle than manganese under these conditions [38].

In addition to organic compounds, *Shewanella putrefaciens* MR-1 is also able to use H_2 as an electron donor for the direct reduction of Mn(IV) [43]. The reaction provided energy for cell growth producing an average of 3×10^6 cells per μmol of Mn(IV) reduced.

3.2 Indirect Reduction

The mechanisms of manganese reduction vary. Reduction may be indirect, in which case physical contact of the microorganism with the mineral particle is not necessary. For example, manganese reduction has been cited to take place through the production and subsequent abiotic activity of microbial metabolites [9, 10].

It remains difficult to differentiate conclusively between direct and indirect methods of microbial manganese reduction. Experiments which can be conducted to distinguish between the two include the ability of spent cell-free media to reduce MnO_2 and the inhibition of reductive dissolution of the manganese mineral oxide by a physical barrier such as dialysis tubing [7, 9].

The bacterium *Achromobacter delicatulus* 182-A excretes citric and malic acids which were found able to dissolve several forms of manganese oxide suggesting that the dissolution of the manganese minerals could be carried out separately from the bacterial cell culture [44].

Di-Ruggiero and Gounot [9] described a bacterium, *Acinetobacter johnsonii*, that reduced manganese via an uncharacterized diffusible product. The dissolution of Mn(IV) by organic microbial metabolites such as formate [28], oxalate, pyruvate, and salicylate has been described [45, 46].

Unlike bacteria, which are procaryotes, fungi are eucaryotic cells and thus are unable to substitute manganese as an electron acceptor for oxygen [28, 47]. Therefore fungi can reduce Mn(IV) only through indirect means. Gupta and Ehrlich [10] described a *Penicillium* sp. that reduced manganese indirectly via an extracellular metabolite which was not identified. However, some strains of *Aspergillus niger* and *Penicillium* sp. are known to produce oxalic acid as a metabolic endproduct of glucose metabolism and thereby reduce Mn(IV) [28, 47].

Oxalate may reduce Mn(IV) as shown in Eq. (2):

$$4H^+ + {}^-OOCCOO^- + MnO_2 = Mn^{2+} + 2CO_2 + 2H_2O \ . \tag{2}$$

Indirect Mn(IV) reduction may also occur through the production of inorganic compounds such as ferrous iron [15, 48], sulfide [6], or other reduced compounds formed during anaerobic respiration or through the production of hydrogen peroxide formed during aerobic respiration [49].

Under anaerobic conditions, Fe^{2+} reduces manganese oxides as follows:

$$4Fe^{2+} + 3MnO_2 + H_2 = 2Fe_2O_3 + 3Mn^{2+} + 2H^+ \ . \tag{3}$$

Sulfide also readily reduces MnO_2 according to the following reaction:

$$MnO_2 + H_2S + 2H^+ = Mn^{2+} + 2H_2O + S^\circ \ . \tag{4}$$

Hydrogen peroxide is another end product of microbial metabolism that can reduce manganese oxide as shown below:

$$H_2O_2 + MnO_2 + 2H^+ = Mn^{2+} + 2H_2O + O_2 \ . \tag{5}$$

Manganese solubilization can also be achieved indirectly by the action of *T. ferrooxidans*. Oxidative leaching of chalcocite or covellite in 9K medium [50] without $FeSO_4$ resulted in the reduction and solubilization of MnO_2 when it was present [51]. Postulated reactions for the leaching of Mn(IV) in the presence of chalcocite are as follows:

$$Cu_2S + MnO_2 + H_2SO_4 = CuS + Cu(OH)_2 + MnSO_4 \tag{6a}$$

$$2Cu_2S + MnO_2 + 1/2O_2 + H_2O + H_2SO_4 = 2CuS + 2Cu(OH)_2 \\ + MnSO_4 \ . \tag{6b}$$

Postulated reactions for the leaching of Mn(IV) in the presence of covellite are as follows:

$$4CuS + 4MnO_2 + 6O_2 + 2H_2O + H_2SO_4 = Cu_4SO_4(OH)_6 \\ + 4MnSO_4 \tag{7a}$$

$$4CuS + 6MnO_2 + 5O_2 + 3H_2SO_4 = Cu_4SO_4(OH)_6 + 6MnSO_4 \ . \tag{7b}$$

Equations (6a) – (7b) indicate the reduction of Mn(IV) coupled with the oxidation of sulfide. Hydroxides of copper and solubilized manganese were not detected in uninoculated controls so the above reactions were believed to be catalyzed by *Thiobacillus ferrooxidans*. The authors predicted that increasing concentrations of manganese dioxide would favor the above reactions resulting in decreased soluble copper.

In other tests, Ghosh and Imai [52] inoculated 9K medium (containing 0.01% $Fe_2(SO)_4$ and 1% elemental sulfur) with *T. ferrooxidans*. The solution was amended with manganese dioxide. During the oxidation of the elemental sulfur, Mn(IV) was reduced and solubilized as $MnSO_4$. Very little manganese was solubilized and bacterial growth was not detected in the absence of elemental sulfur.

When manganese is in the form of carbonates or silicates, microbial leaching may occur due to microbial acid or ligand production. A reaction for the acid dissolution of rhodochrosite ($MnCO_3$) is as follows [2]:

$$MnCO_3 + 2H^+ = Mn^{2+} + H_2O + CO_2 . \tag{8}$$

4 Chemical Manganese Reduction

Stone and Morgan [45] found that several organic compounds could solubilize manganese under abiotic conditions including methoxycatechol, dihydroxybenzoic acid, ascorbate, nitrocatechol, thiosalicylate, hydroquinone, syringic acid, *o*-methoxyphenol, vanillic acid, orcinol, and resorcinol. Troshanov [53] showed that reducing sugars such as glucose and xylose could also reduce Mn(IV) under abiotic conditions. Noble et al. [54] tested nine organic acids for the ability to leach manganese from ore (Table 1). Although these acids were not tested using

Table 1. Abiotic leaching of Three Kids ore with organic acids. Conditions: 2 g ore, 100 ml solution containing 4 g l^{-1} organic carbon, ambient temperature, 200 rpm

Acid	% Manganese Extraction	Days to Max. Leaching
L-Malic	91	7
α-Ketoglutaric	91	8
Citric	84	3
Formic	77	15
Lactic	56	11
Oxalic	16	2
Succinic	6	15
Fumaric	4	10
Acetic	4	8

abiotic conditions, they found that rates of leaching were similar whether or not sodium azide was present. Ehrlich et al. [55] also showed that glucose can reduce Mn(IV) under sterile conditions but that the addition of the manganese-reducing bacterium *Bacillus* GJ 33 significantly accelerated the dissolution process.

5 Mining Applications for Microbial Manganese Leaching

5.1 Recovery of Manganese

Manganese-reducing bacteria have been used to extract manganese from manganiferous ores. Both pure and mixed cultures have been tested on a laboratory scale.

Noble et al. [54] have investigated bioleach processes to extract manganese from low grade ores using shake flask, column tests, and heap tests. The initial medium used was a glucose-mineral salts medium. The authors found that the exclusion of phosphate from the medium inhibited the growth of fungi thereby enhancing the leaching process. The fungi caused depletion of carbon with a concomitant increase in pH and precipitation of the solubilized manganese.

Noble et al. [54] conducted leach tests in 50-mm columns using ore from the Three Kids deposit in Clark County, Nevada ground to − 6.3 mm. The first tests were run with the glucose-mineral salts medium. Occasional replacement of the mineral salts medium or glucose addition resulted in jumps in leach rates. Up to 28% of the manganese was extracted in 52 weeks. Later column tests were conducted using 3% molasses. Once again, media changes accelerated the rate of manganese extraction (Fig. 2). In this case, 99% of the manganese was solubilized in 52 weeks.

Noble et al. [54] also ran laboratory-scale heap leach tests by bioleaching 35 kg of − 19 mm ore in 33 cm diameter containers. Over a year (57 weeks) was required to achieve a 10% extraction of manganese. Further work was planned to improve the heap leach results.

Madgwick [12] conducted laboratory-scale studies to examine the microbial dissolution of manganese from Groote Eylandt tailings. The Groote Eylandt deposit is in the Northern Territory of Australia. The waste tailings contain approximately 17% Mn. A commercial process needed to be developed which yielded $3-6\,\mathrm{g\,Mn^{2+}\,l^{-1}}$ in the pregnant solution. Bioleach tests were run using molasses-based growth media in stationary flask tests and in stirred 5 l bioreactors. In the former case, average rates of manganese extraction were $0.8\,\mathrm{g\,Mn^{2+}\,l^{-1}d^{-1}}$. In the latter case, an average extraction rate of $1.1\,\mathrm{g}$ $\mathrm{Mn^{2+}\,l^{-1}d^{-1}}$ was achieved.

Srimekanond et al. [56] used a mixture of *Enterobacter cloacae* and *E. agglomerans* to extract manganese from ore from the Groote Eylandt deposit. This deposit is of Cretaceous origin and contains mixtures of cryptomelane, pyro-

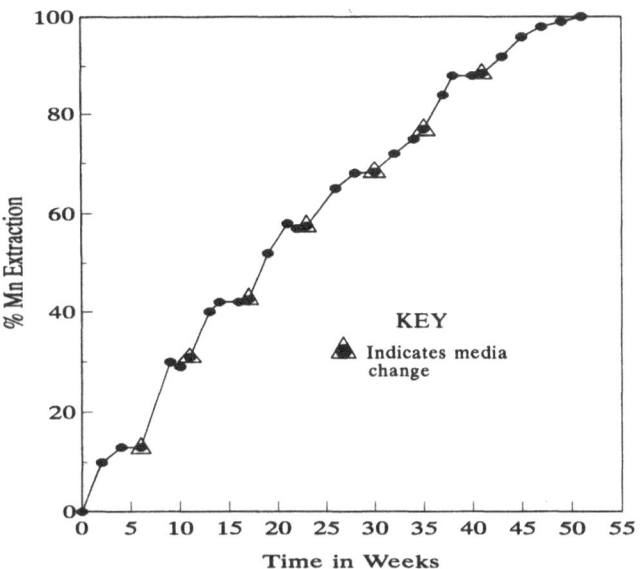

Fig. 2. Column leaching of Three Kids ore with 3 food-grade molasses

lusite, romanechite, todorokite, and gangue minerals including goethite. The manganese minerals were solubilized in the following order: cryptomelane/todorokite, romanechite, and pyrolusite, demonstrating that the mineral form affected the rate of dissolution.

These investigators used a repeated static batch process in which a 20% pulp density of ore was mixed with a nonsterile growth medium consisting of 5% molasses. Incubation was carried out under static conditions at 32 °C with a single resuspension of particles by mechanical agitation once a day. One hundred ml volumes of culture were incubated in 500 ml conical flasks loosely plugged with cotton wool. At each 5 day interval, a solution/solids separation was performed and the adsorbed manganous ion was desorbed from the oxide with $0.03 \ mol \, l^{-1}$ sulfuric acid. The acid washings were combined and released manganous ion termed 'acid desorbed Mn^{2+}.' The manganous ions in the supernatant were called 'soluble Mn^{2+}.' The solids were neutralized by consecutive washings in reverse osmosis water and returned for the next incubation with fresh inocula and new nutrient media.

Two ore types were tested – a cryptomelane-rich ore and a pyrolusite-rich ore. Biological extraction yielded 63% and 53% soluble manganese over a 71 day period from the cryptomelane- and pyrolusite-rich samples, respectively. In either case, the pH did not drop below 4.9. However, much of the solubilized manganese had been adsorbed to the oxide mineral surface.

Mercz and Madgwick [24] used a 2% pulp density in a mineral salts medium with glucose. The ore was also a pyrolusite ore from the Groote Eylandt deposit ground to −74 µm. The ore suspension was autoclaved, inoculated and incubated at 30 °C in 250-ml Erlenmeyer flasks. After solid/solution separations, the soluble manganese in the supernatants was determined by atomic absorption spectroscopy (AA). In some cases, algal cultures were added to the growth media with unsterile ore and inoculated with a mixed slime-dam bacterial culture. Algal extracts used included cell-free spent media, crude extracts, and purified carbohydrate extracts. Using arabinose and mannitol as the carbohydrate sources in the growth medium resulted in the greatest extraction of manganese whether mixed or pure cultures of bacteria were used. Whole algal cultures, without the addition of a carbohydrate, resulted in only marginal release of manganese. Crude algal extracts and cell-free spent media did not enhance microbial manganese solubilization. This may have been due to the effect of pH and/or the presence of inhibitory algal products. However, purified algal extract did enhance manganese dissolution by a pure culture of bacterial isolate I-29S.

Marine ferromanganese nodules also represent a significant reservoir of manganese. Ehrlich et al. [55] used nodules that contained approximately (in percent dry weight) 22.8–20.6% Mn, 6.6–6.5% Fe, 1.1% Ni, 0.7% Cu, and 0.2–0.3% Co. Most of the iron and manganese are in the oxidized form [57, 58]. Copper and nickel are thought to be primarily associated with the manganese minerals and the cobalt incorporated into the lattice structure of the iron minerals [59]. When marine manganese nodules are biosolubilized, cobalt, copper, and nickel are dissolved as well as manganese [2, 55, 60]. The dissolution of nickel followed that of manganese which indicated that nickel may have been adsorbed to the manganese oxide mineral and was released as the mineral dissolved [55]. Cobalt may have been released in manner similar to that of nickel or the Co(III) may have been biologically reduced to Co(II). Copper was also bound to manganese oxide surfaces in the nodules, although more loosely than nickel. Negligible amounts of iron were solubilized. The lack of iron dissolution may have been because the iron was in a form in the nodule matrix not readily accessible to the enzyme.

5.2 Separation of Manganese from Iron

Manganese-reducing bacteria may also be used to separate manganese from iron in ferromanganiferous ores. Ore from an Arizona deposit designated AZ-1 was bioleached in 2 l bioreactors [61]. The ore contained 5.8% Mn and 12.0% Fe. Three hundred-gram samples of nonsterile AZ-1 ore were added to 2 l bioreactors with 2000 ml of an active culture of *Bacillus* MBX 1 in PGM medium (proprietary). Anaerobic conditions were maintained by continuously sparging the reactor with nitrogen. Reactor contents were stirred at 150 rpm and the temperature was maintained at 35 °C. The experiment continued for 14 days. On a daily basis, half of the reactor contents were siphoned, centrifuged, and the supernatant removed. Solubilized metals were analyzed by atomic absorption. The solids were returned to the bioreactor along with 1 l of fresh bacterial inoculum. Planktonic cell numbers of native populations of bacteria were monitored in addition to cell numbers of *Bacillus* MBX 1. The solution pH was monitored daily and adjusted as needed to maintain a range of 6.5 to 8.5.

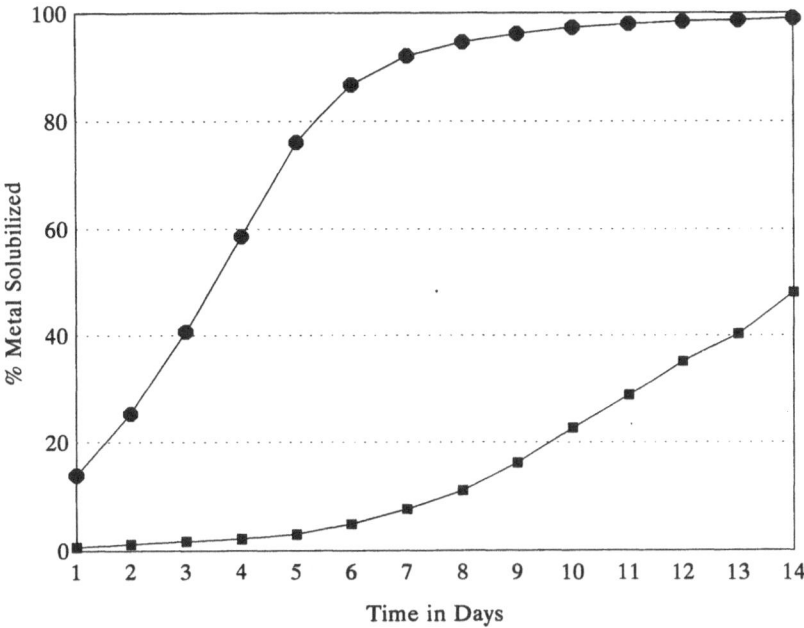

Fig. 3. Microbial solubilization of manganese and iron from ore in 21 bioreactor. ● = manganese; ■ = iron

At the end of the experiment, the biotreated tails were washed three times by centrifugation with distilled water, dried, weighed, digested by aqua regia, and analyzed for residual metals.

As shown in Fig. 3, most of the manganese solubilization occurred by day 6. As manganic manganese availability decreased, ferric iron solubilization began to increase, suggesting that Fe(III) was acting as a final electron acceptor only after Mn(IV) was no longer available. The order of manganese and iron reduction might be predicted based on the standard redox potentials of the two metals [62]:

$$E_H{}^\circ$$

$$\tfrac{1}{2} \, MnO_2 + 2 \, H^+ + e^- = \tfrac{1}{2} \, Mn^{2+} + H_2O \quad +1.29 \tag{9}$$

$$FeOOH + 3 \, H^+ + e^- = Fe^{2+} + 2 \, H_2O \quad +0.67 \,. \tag{10}$$

However, the apparent step-wise solubilization of manganese and iron may have been due to the oxidation of Fe(II) by Mn(IV) rather than to the difference in redox potentials. The order in which metals are reduced may also depend on the relative abundance and availability of different electron acceptors, the affinity of different organism's respiratory enzymes for available electron donors, abiotic reactions between the respiratory product of one organism and the electron acceptor of another, and other factors [63].

Cell numbers of isolate *Bacillus* MBX 1 remained high (10^7 ml^{-1}) despite greater numbers of native bacteria (10^9 ml^{-1}). Survival of the inoculated bacteria

Table 2. AZ-1 bioreactor study: percent metal biosolubilization by *Bacillus* MBX 1

	Mn	Pb	Ag	Fe	Cd
Bioreactor A	98.9	80.3	36.1	36.5	74.6
Bioreactor B	99.1	81.8	30.7	48.9	80.3
Average	99.0	81.1	33.4	42.7	77.5

in the presence of environmental competitors is important since sterilization of tailings in the field would be impractical.

Lead, silver, and cadmium were also extracted during the bioleach of ore AZ-1 (Table 2). The silver was leached from sites in the pyrolusite, where it had been occluded, as the crystalline structure was dissolved. Lead and cadmium do not have multiple positive valences. Therefore, the dissolution of these metals was not due to direct microbial reduction. Nor were these metals thought to be encapsulated by the pyrolusite mineral. The solubilization of lead and cadmium was probably due to the production of organic acids by *Bacillus* MBX 1 including acetic acid.

5.3 Recovery of Silver from Refractory Manganiferous Ore

Manganese-reducing microorganisms have been used to liberate silver encapsulated by manganese oxides. One manganiferous silver ore tested was obtained from a deposit in Colorado. The Crystal Hill mine site in Saguochi County Colorado included milled ore that had been treated with cyanide during gold leaching operations. The leachate from the ore heaps was collected in pregnant and barren ponds which still existed on the property. The manganese in the ore was primarily in the form of pyrolusite containing residual refractory (to cyanide-leaching) silver. Manganese-reducing bacteria broke down the crystalline structure of pyrolusite rendering the silver more amenable to further cyanide extraction [19].

The most extensively tested refractory manganiferous silver oxide ore tested was from Santa Cruz County in Arizona [13, 64]. This deposit is one of the larger silver deposits in the United States with an excess of 5.4×10^9 kg of near-surface mineralization containing about 170 $\mu g\,g^{-1}$ silver [3]. This deposit formed as the result of oxidation of a manganese sulfide deposit which had previously replaced permeable Cretaceous ash-flow volcanics and Paleozoic sediments. The deposit contains members of the Cryptomelane-Coronadite group, (K, Pb, Zn, Cu, Ag, Ba) $(Mn^{+2}, Mn^{+4})_8 O_{16}$. Cations, including Ag^{+1}, are contained in the tunnel site of the structure, XMn_8O_{16} rendering the silver resistant to cyanide leaching. These manganese oxide minerals must be reduced to liberate the entrapped silver and allow it to be contacted with leach solutions. Baseline 0.1% cyanide bottle roll tests resulted in only 13.7% extraction of the silver. Chemical agents such as sulfur dioxide which can be used to reduce the Mn(IV) had become environmentally unacceptable. Metal-reducing bacteria offered an alternative to the use of harsh chemicals for the treatment of these ores.

Bioreactor studies were performed with the Arizona manganiferous oxide silver ore. Fourteen-day bioreductive leach tests were conducted in 2-l bioreactors. The ore was slurried with cultures of *Bacillus* MBX 1 at 15% pulp density using -200 mesh particle size. Bioreactor A was supplemented with $19.1 \, g \, l^{-1}$ organic collector while Bioreactor B was supplemented with half the concentration of collector. The purpose of the collector was to aid in the stabilization of solubilized metals. Over 86% of the silver was biosolubilized into the organic growth medium in Bioreactor A by *Bacillus* MBX 1. In addition, 99.8% of the manganese, $> 99\%$ of the copper, and 91% of the zinc in the ore were solubilized into the growth medium. A subsequent 3 day cyanide bottle roll of the residual solids from Bioreactor A solubilized an additional 8.5% of the silver for a total extraction of 94.5%. The importance of the collector was underscored by the fact that only 45.0% of the silver was biosolubilized into the growth medium in Bioreactor B with 45.5% additional dissolution during bottle roll leaching of the biotreated solids for a total combined extraction of 90.5%. Bacterial leaching was also critical to this success as evidenced by the fact that the total combined solubilization of silver in the sterile uninoculated control sample was only 10%.

These results indicated that cyanide was not required for extraction of large quantities of silver from this refractory ore deposit. These tests also showed that manganese-reducing bacteria can act as powerful catalysts for the reduction of manganese. Indeed, a dramatic color change was seen in the ore during the bioleaching period. This was primarily due to the bacterial reduction and solubilization of 99.8% of the manganese from the ore sample in Bioreactor A.

The solubilization of copper and zinc was probably not due to bacterial reduction. For example, in all its compounds, zinc shows only a $+2$ oxidation state. Reduction of zinc to its elemental form would not result in its solubilization. However, many organic acids have a strong affinity for copper and zinc. Some of the organic acids that bacteria secrete, including acetic and lactic acids, have the ability to chelate heavy metals including zinc and copper. Thus, the dissolution of these metals was most likely due to organic acid production by the *Bacillus* during the bioleaching process.

6 Optimal Industrial Processes for Microbial Manganese Solubilization

6.1 Growth Substrate

In many cases, identification of a manganese-reducing bacterium will tell a great deal about its survival and growth requirements and will aid in tracking the organism during the manganese dissolution process. Bacterial isolates can be identified according to Bergey's Manual [65]. Gram reactions are determined, motility or lack thereof is established, and spores identified. Additional important tests include catalase and oxidase tests as well as starch and gelatin hydrolysis. Biochemical characteristics can be determined using various packaged systems such as the Micro-ID test system (Organon Teknika Corp., Durham, NC) Biolog, (Hayward, CA), and the API system (bioMerieux, Hazelwood, MI).

The bacteria described in the literature that directly reduce manganese are heterotrophs, i.e., they require a source of organic carbon. Other important elements include nitrogen, phosphorus, and sulfur. (Many other elements are necessary in lesser amounts.) When a complex growth medium is used in the manganese extraction process, such as molasses, whey, or PGM, all the growth requirements are likely to be satisfied. However, if the operator plans to use a defined mineral salts medium, growth factor requirements should also be determined [11].

Cheap nutrient sources include molasses, corn starch hydrolysate, beet pulp, digested sewage sludge, whey, brewers yeast waste (first pass beer [66]), and ammonium lignosulfate (a paper industry byproduct [67]). Molasses, the most commonly used growth substrate, may contain 35% sucrose, 1–12% glucose, fructose, starch and other minor carbohydrates, 7–15% ash (K, Mg, Ca, Na, SO_4, Cl, P and Si), 2–5% nitrogenous compounds, 2–8% organic acids, and vitamins A and B [56, 68]. Potassium accounts for 30–50% of the inorganic elements in the ash [56].

After choosing an organic carbon and energy source, it is important to determine the optimum concentration for manganese leaching. Silverio and Madgwick [69] found that the optimal rate of leaching of manganese dioxide tailings occurred using 5% sucrose. These authors thought that the higher concentrations of sucrose depressed leaching due to the osmotic stresses of a hypertonic solution.

Certain forms of nutrients may depress the extraction rates of manganese. For example, Veglio et al. [70] conducted aerobic shake flask experiments for the microbial solubilization of manganese oxide in ore samples obtained from the Casale Castiglone area in Italy. The presence of sucrose and ammonium nitrate in the growth medium had a positive effect on manganese dissolution. However, dihydrogen potassium phosphate had a negative effect. The effects of manganese sulfate and calcium chloride were negligible.

Kozub and Madgwick [68] found that algal supplementation markedly enhanced manganese dissolution from a Groote Eylandt manganese oxide ore in the presence of all sugars tested. Dried algal cells were added at a concentration of 0.03% dry weight while sugars were added at a 1% concentration. Without algal supplementation the most effective sugars (in terms of manganese dissolution) showed the following ranking: fructose > galactose > sucrose. With algal supplementation the ranking of the sugars was reversed. The effect of algal supplementation with molasses as the carbohydrate source was equivocal.

Hart and Madgwick [71] found that a cheap alternative to molasses was the dried powder of the red alga *Gracilaria secundata* but the dried seaweed was only half as effective as molasses on a weight-to-weight basis. Madgwick [12] found that the addition of peptone, yeast extract, ammonium sulfate, ammonium nitrate, or ammonium phosphate depressed manganese leaching (concentrations used not given). Madgwick felt this was because the added nitrogen stimulated the growth of non-leaching bacteria over the manganese-leaching bacteria.

According to Madgwick [12], the efficiency of manganese dissolution per gram of carbohydrate varies with microbial species, concentration of carbon source, and the pulp density of the manganese oxide ore. At higher than necessary concentrations of carbohydrate, the efficiency of manganese reduction diminishes due to the growth of bacteria that are not actively leaching and metabolic waste of nutrient when a surface dependent reaction has become saturated.

6.2 Electron Donor

Very often, the carbon source for heterotrophic bacteria also serves as an energy source, i.e., it also serves as a source of electrons. Microbial cells, like all living cells, derive energy from a carefully orchestrated organized flow of electrons through a chain of organic electron carriers. The ultimate transfer of these electrons to a terminal electron acceptor by this respiratory system enables the cell to conserve biochemical energy. Sources of organic carbon such as glucose, and many other sugars, also serve as electron donors. Many constituents in complex organic growth media will serve the dual roles of nutrient and energy source.

A dilute mineral salts medium can also be used for the microbial solubilization of manganese so long as the bacteria have adequate sources of carbon and energy. However, the organic growth medium used at MBX, (PGM), offers several advantages over a mineral salts medium. The bacteria grow more rapidly, the pH remains stable, and the extraction rate of target metals is greater than when using a defined mineral salts medium [64].

6.3 Atmospheric Oxygen

In some cases, microbial reduction of manganese requires anaerobic conditions. Previous studies [7, 8, 31] have shown that manganese reduction may be inhibited by molecular oxygen and takes place only under anaerobic conditions. Some of the bacteria that reduce manganese only under anaerobic conditions include *Geobacter metallireducans* GS-15 [7], *Shewanella putrefaciens* [8], and *Bacillus polymyxa* MBX 2 [11]. *Geobacter metallireducans* GS-15 is a strict anaerobe, unable to grow in the presence of oxygen. *Shewanella putrefaciens* is aerobic but is able to survive without oxygen by using anaerobic respiration if suitable electron acceptors, such as Mn(IV), are present. *Bacillus polymyxa* MBX 2 can switch from aerobic to anaerobic metabolism if deprived of a source of oxygen. Under anaerobic conditions it remains unclear whether Mn(IV) is reduced by a process of anaerobic respiration by attached cells or whether the Mn(IV) is used as an electron sink for the reoxidation of NADH.

Some bacteria are able to reduce manganese at similar rates anaerobically or aerobically such as *Bacillus* 29, although this bacterium requires aerobic conditions to be induced to reduce Mn(IV) [39].

In general, the most rapid rates of manganese dissolution are achieved under microaerobic or anaerobic conditions. Kozub and Madgwick [68] found a three- to ten-fold increase in bacterial manganese dissolution using microaerobic rather than aerobic conditions. Later studies by Madgwick [12] confirmed this, showing that a five-fold increase in manganese dissolution occurred under microaerobic conditions as compared to aerobic cultures. There was also a significant increase in manganese solubilization per gram of carbohydrate consumed.

Under anaerobic conditions, Mn(IV) or Mn(III) probably act as an electron acceptor and direct interaction between the bacterium and the mineral surface is necessary [12]. Organic acids are probably the primary reductant of manganese under aerobic conditions as manganese dissolution occurs via diffusible low molecular weight organic reducing compounds.

6.4 Temperature and pH

Process temperature is also important to the microbial dissolution of manganese. Many manganese-reducing bacteria described in the literature are mesophilic with an optimum growth range of 25 to 35 °C. The enzymes in these bacteria generally act more rapidly at higher temperatures, doubling their reaction rate for every 10 °C increase. Therefore, the enzymatic reduction of manganese will occur more rapidly at higher temperatures than at lower ones. The limit on the upper end of the temperature scale is imposed by the fact that enzymes are proteins which become inactivated (or denatured) at excessively high temperatures. A manganese-solubilizing bacterium's optimum temperature will be near its high temperature limit. Therefore, careful temperature control must be maintained to achieve optimal rates of microbial manganese dissolution without injury to the microorganisms.

The optimum temperature range will need to be determined on a case by case basis depending upon the microbial population used. Srimekanond et al. [72] found, in their study, that the most efficient microbial manganese dissolution occurred at 50 °C with the least leaching activity at 4 °C and 20 °C. A temperature range of 4 to 70 °C was tested. On average, the rate of manganese solubilization at 50 °C was 2.8 times that at 70 °C. In contrast, Hart and Madgwick [73] found that the optimum temperature of incubation for the microbial solubilization of manganese in Groote Eylandt ore was 30–35 °C. The rate of leaching increased five-fold between 20 °C and 30 °C.

The effect of pH on microbial growth and leaching rates must also be determined. Most manganese-solubilizing bacteria are neutrophilic with an optimum pH range between 6 and 8. However, the pH tolerance range may be much greater than the optimum range. Methods to determine the optimum pH range are discussed by Rusin et al. [11].

Process pH will affect manganese chemistry as well as the microorganisms. Manganous ions tend to adsorb to MnO_2 at pH > 5 which limits further leaching [74]. However, unlike Fe(II), Mn(II) does not readily autooxidize at pH values below approximately 8.0. In order to accelerate the leaching process, metal ions may be desorbed from mineral particles using dilute sulfuric acid [75] or chelators [76].

The pH produced in the culture medium may be affected by the type of manganese mineral under leach. Serebrjanaja et al. [44] found that the solution pH after 10 days of bioleach was 2.8 and 4.5 when leaching brownite and rancieite respectively. The optimum pH range for *Achromobacter delicatulus* 182-A, used in this study, was 5.5 to 6.5. At lower pH values, microbial growth ceased. In this case, pH control would be necessary. However, Hart and Madgwick [73] found that the regulation of pH in batch leaching of manganese dioxide tailings containing cryptomelane, pyrolusite, romanechite, and todorokite in molasses growth media was unnecessary. The pH self-adjusted to pH 5.8 ± 0.6 after one week of incubation.

6.5 *Metal Toxicity*

Metal toxicity is an important consideration for any microbial metal dissolution process. Most of the metals in ores are present in relatively insoluble forms. Metals are toxic to bacteria only in soluble ionic forms; therefore, mineralogical analyses of ores are not always reliable predictors of bacterial inhibition. Heavy metal toxicity tests can be conducted for manganese-reducing bacteria. However, the results are valid only for the conditions under which the testing is performed. Resistance to heavy metals for any one bacterial strain will differ depending on parameters such as pH, Eh, the presence of other cations and anions, the concentration of organic matter, and temperature [77]. By controlling these parameters, we can manipulate bacterial metal sensitivities to some degree. Even so, high concentrations of soluble metals may be toxic.

6.6 *Culture Process*

The type of manganese oxide mineral under leach will influence retention time and the economic feasibility of using biological methods. Serebrjanaja et al. [44] found that different forms of manganese minerals leached at different rates using *Achromobacter delicatulus* 182-A. The leaching rate in descending order was rancieite = todorokite > cryptomelane = pyrolusite > manganite > ramsdellite > brownite. No connection could be made between the chemical composition of the minerals and the rate of biological leaching. These investigators found that mixtures of manganese minerals leached more rapidly than when purified minerals were tested. They suggested that it may be possible to optimize rates of leaching by using mixtures of ores with different mineral compositions.

Rusin et al. [64] also found that retention time depended partly on the mineralogy of the individual ore. Microbial manganese leaching was conducted in order to extract silver from pyrolusite crystals. Silver was bioleached from a Colorado manganiferous silver ore much more rapidly than from an Arizona ore under the same conditions due to the differences in the manganese/silver ratios. The Mn/Ag ratio in the Colorado ore was 104.1 while the Mn/Ag ratio in the Arizona ore was 742.6. The bacteria were required to solubilize far more manganese to free equivalent amounts of silver in the Arizona ore than in the Colorado ore.

The use of a bioreactor enables the operator to maintain more control over the microbial dissolution process. It has been shown, at the laboratory level, that elimination of the indigenous bacteria is not necessary before inoculating a bioreactor with manganese-reducing bacteria for the extraction of manganese from ore. The presence of native bacteria in the bioreactors did not appear to inhibit the dissolution of metals in ores by manganese-reducing inocula [64].

Silverio and Madgwick [69] also found that the presence of non-leaching native bacteria was not necessarily a drawback to the extraction of manganese from low grade ores or tailings. They showed that the use of crude enrichment cultures resulted in higher rates of manganese extraction from tailings than the use of purified manganese-reducing cultures. However, the most efficient extraction, based on grams of Mn^{2+} extracted per gram sucrose consumed, was achieved by them using a pure culture of Mn(IV)-reducing *Achromobacter*.

It might seem that each ore deposit would contain its own native bacteria which would be best suited to reduce manganese in that ore. However, a bioreactor represents a foreign environment, quite different from the natural source. Therefore the most efficient metal-solubilizing bacteria in a bioreactor are not necessarily indigenous strains. This possibility, suggested by Holmes in 1988 [78], was confirmed by Rusin et al. [64]. Although many manganese-solubilizing bacteria were isolated from various ore samples, the most rapid manganese-reducing strains for a particular ore were not always the indigenous isolates. For instance, the most efficient manganese-reducing bacterium for a Colorado ore was *Bacillus* MBX 2, although it was originally isolated from an Arizona deposit. However, *Bacillus* MBX 1 was one of the most rapid manganese-reducing bacteria when tested with its native Arizona ore.

If a bioreactor is inoculated with manganese-solubilizing bacteria, the kinetics of manganese solubilization by the chosen bacterium should be a consideration. Rates of manganese reduction were compared between *Bacillus polymyxa* MBX 2, *Shewanella putrefaciens* sp 200, and *Shewanella putrefaciens* MR-1 under anaerobic conditions [11]. The electron donor for isolate MBX 2 was 10 mmol1^{-1} glucose while the electron donor for MR-1 and sp 200 was 10 mmol1^{-1} lactate. Isolate MBX 2 reduced almost three times the amount of manganese reduced by MR-1 in 250 h and five times that of sp 200 (Fig. 4). Strain MBX 2 is among the most rapid Mn(IV)-reducing bacteria described to date [11].

The use of an organic culture medium is subject to microbial contamination. If desired, contamination by non-leaching heterotrophs can be minimized by using

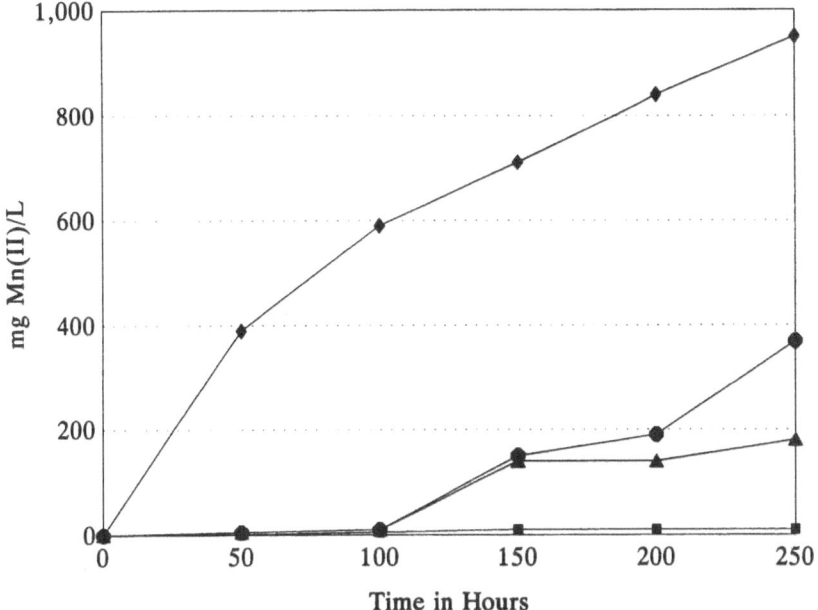

Fig. 4. Comparison of microbial rates of manganese dissolution. ♦ = *Bacillus polymyxa* MBX 2; ● = *Shewanella putrefaciens* MR-1; ▲ = *S. putrefaciens* sp 200; ■ = sterile control

selective culture conditions such as acid pH, anaerobiosis, a specialized carbon and/or energy source and the presence of inhibitory agents such as salt (if using a marine Mn(IV)-reducing bacterium). Such selective pressures may enable the operator to maximize conditions for a target microbial population. In some cases, a pure manganese-reducing culture may be introduced into a bioreactor, but the use of mixed cultures may lead to better results.

Srimekanond et al. [56] reported that a mixture of manganese-reducing *E. cloacae* and *E. agglomerans* resulted in a maximum rate of manganese leaching four to six times greater than the rate achieved using either organism in pure culture. Holden and Madgwick [79] also found that a mixed culture was more beneficial than a pure isolate. When using a mixture of *Pseudomonas* and *Bacillus*, the *Pseudomonas* seemed to enhance glucose and nitrogen utilization by the *Bacillus*.

It is quite probable that manganese-reducing bacteria adsorb to the oxide mineral surfaces in a manner similar to that described for *Thiobacillus* and sulfur particles [80]. Indeed, the work of Ghiorse and Ehrlich [29] had already suggested this to be the case. Madgwick [12] found that when manganese-reducing bacteria were allowed direct access to manganese oxide mineral particles by removing a permeable barrier, there was an eight-fold increase in the rate of manganese dissolution. Therefore, more rapid dissolution of manganese will occur if bacteria are in direct contact with the ore rather than when using spent growth media.

The tendency of bacteria to form biofilms on particles may contribute to their ability to act as efficient living chemical reagents. Whether bacteria solubilize a metal via chemical reduction or chelation with organic acids, direct point contact between the bacterial cell and the mineral particle will optimize the dissolution process. Attachment precludes the necessity for adding an intermediate (soluble) electron carrier in the case of microbial reduction. If dissolution is due to the extracellular production of a metabolite such as formic acid, direct contact between mineral and bacterium will minimize diffusion of the reactive chemical from the mineral surface. Therefore, the optimal ratio of cell numbers to mineral surface area will be an important consideration in the design of a bioreactor system.

One of the advantages that the use of a bioreactor offers over that of heap leaching is agitation of the ore slurry during the leach process. Hart and Madgwick [73] found that the overall performance of manganese leaching in stirred reactor tanks was superior to predominantly stationary conditions. The Mn(II) yields per gram carbohydrate used and rates of manganese dissolution were improved by mixing the ore slurry. Ehrlich [2] also found that agitation increased microbial manganese dissolution under aerobic or anaerobic conditions. The highest percent Mn(IV) solubilization was observed under anaerobic conditions at the highest agitation rate (300 rpm). The beneficial effect of shaking may be due to better contact of the culture medium with bacteria and mineral surfaces for the exchange of reactants and products of the reaction.

Pulp density will also be an important factor. At a laboratory scale, Buys et al. [75] found that the maximum rates of manganese dissolution varied with the concentration of molasses in the feed solution and pulp density. Overall, the most rapid rates of manganese extraction occurred using a 14% pulp density (w/v). The highest percent extraction was also achieved at the same pulp density.

If heap or dump leaching is employed to extract manganese, inoculation with manganese-reducing bacteria is likely to be ineffective due to competition by indigenous organisms. In addition, it is more difficult to control process parameters such as temperature and pH. However, capital and operating costs are much lower for heap leaching than for running bioreactors. When comparing costs, however, one should remember that a bioreactor, in this case, would not require aeration. Anaerobiosis will not be difficult to achieve as facultative bacteria will rapidly scavenge the available oxygen [31].

In either case, soluble manganese can be recovered from the pregnant solution by raising the pH with lime, solvent extraction, or electrolytically. The choice of the recovery method will depend on the composition of the pregnant solution and the concentration of manganese.

7 Summary

Manganese-reducing microorganisms can be isolated from various habitats using established methods. Most manganese-reducing bacteria require direct contact with ore particles to achieve manganese dissolution while fungi solubilize Mn(IV) indirectly via extracellular metabolites.

Manganese-reducing microorganisms can be used to extract manganese from low-grade ores and tailings. Suitable deposits are found world-wide. These organisms can also be used to separate Mn from Fe in iron ores. In addition, refractory manganiferous silver ores in which the target metal is entrapped by oxide minerals can be bioleached with Mn(IV)-reductive bacteria with extractions of 86–93%. Metal extraction can be achieved through bacterial leaching without the use of cyanide. The primary end-product of this bioleaching process is spent bacterial growth medium of a neutral pH which may enhance plant nutrition in the remediation of tailings.

7.1 Future Applications

Future applications of microbial manganese leaching probably depend on the use of direct enzymatic reduction processes. Madgwick [12] estimated that the costs of microbial manganese leaching was competitive with that of chemical reduction (no cost data shown).

The rates of microbial manganese solubilization reported thus far are generally slower than those obtainable by hydrometallurgical methods. However, optimization may render the microbial process more attractive. For example, Abbruzzese et al. [81] achieved a 98% extraction of manganese after 20 days leaching with 6 vol.% sulfur dioxide solution. When the authors used biological leaching, complete dissolution of manganese was obtained in 15 days.

The Dean-Leute ammonium carbamate process could be used to extract manganese from low grade ores but it was found to be too expensive [82]. Reduction can also be achieved with sulfur dioxide and acid ferrous sulfate [83]. However, the ferrous sulfate method is corrosive and sulfur dioxide can present disposal problems. Spent pickling solution (1–5% sulfuric acid, 10–25% ferrous sulfate) can also be used but the high ferric iron concentration in the pregnant solution also creates disposal problems [81]. Therefore, bioleaching is an environmentally benign alternative to pyrometallurgical techniques. Although the use of bioreactors allows for optimal control of the leaching process, heap or in-situ leaching will probably be more economically feasible for low-grade manganese resources. These latter techniques offer the advantages of low capital costs, low operating costs, and short start-up times.

The rate of heap leaching of these low-grade ores may not be critical due to the low operating costs. The most important cost factor will be the source of organic carbon for the heterotrophic microorganisms. Many types of food process residues can be used but if the source is too far from the mine site, transportation costs could be prohibitive. Hart and Madgwick [71] have suggested that marine algae can be cultivated on site and used as a source of microbial nutrients.

References

1. Jones TS (1985). In: Mineral facts and problems, Bureau of Mines Bulletin 675, US Department of the Interior, Washington DC, p 483
2. Ehrlich HL (1988). In: Proceedings of the 8th International Biotechnology Symposium, vol 2, Paris, p 1094
3. Koutz FR (1984) The Hardshell silver, base-metal, manganese oxide deposit, Patagonia mountains, Santa Cruz County, Arizona: a field trip guide. The Arizona Geological Society Digest, vol 15, p 199
4. Ehrlich HL (1973). In: Phase I Report, inter-university program of research on ferromanganese deposits on the ocean floor. Seabed assessment program, international decade of ocean exploration. National Science Foundation, Washington DC, p 217
5. Ehrlich HL (1963) Appl Microbiol 2: 15
6. Burdige DJ, Nealson KH (1985) Appl Environ Microbiol 50: 491
7. Lovley DR, Phillips EJP (1988) Appl Environ Microbiol 54: 1472
8. Myers CR, Nealson KH (1988) Science 240: 1319
9. Di-Ruggiero J, Gounot AM (1990) Microb Ecol 20: 53
10. Gupta A, Ehrlich HL (1989) J Biotechnol 9: 287
11. Rusin PA, Quintana L, Sinclair NA, Arnold RG, Oden KL (1991) Geomicrobiol J 9: 13
12. Madgwick JC (1993). In: Torma AE, Wey JE, Lakshmanan VL (eds) Biohydrometallurgical technologies. The Minerals, Metals & Materials Society, vol 1, p 343
13. Rusin P, Cassells J, Sharp J, Arnold R, Sinclair NA (1992) Minerals Eng 5: 1345
14. Ehrlich HL (1966) Dev Ind Microbiol 7: 279
15. Myers CR, Nealson KH (1988) Geochim Cosmochim Acta 52: 2727
16. Paik G (1980). In: Lennette EH (ed-in-chief) Manual of clinical microbiology 3rd edn. American Society for Microbiology, Washington, DC, p 1016
17. American Public Health Association (1989) Clerceri LS, Greenberg AE, Trussell RR (eds) Standard Methods for the Examination of Water and Wastewater. 17th ed. p 9–35
18. Wahab AAM (1975) Plant Soil 42: 703
19. Rusin PA, Sharp JE, Arnold RG, Sinclair NA (1991). In: Smith RW, Mishra M (eds) Mineral bioprocessing. The Minerals, Metals, and Materials Society, New York, p 207
20. Arnold RG, DiChristina TJ, Hoffman MR (1986) Appl Environ Microbiol 52: 281
21. Arnold RG, Hoffmann MR, DiChristina TJ, Picardel FW (1990) Appl Environ Microbiol 56: 2811
22. Ehrlich HL (1993) J Ind Microbiol 12: 121
23. Ehrlich HL (1987) Geomicrobiol J 5: 423
24. Mercz TI, Madgwick JC (1982) Proc Aust Inst Min Metall 283: 43
25. Agate AD, Deshpande HA (1977). In: Schwartz W (ed) Conference – Bacterial Leaching, Verlag Chemie, Weinheim, p 241
26. Babenko YS, Dolgikh LM, Serebryanaya MZ (1983) Mikrobiologiya 52: 674
27. Groudev SN (1987) Acta Biotechnol 4: 299
28. Ehrlich HL (1980). In: Biogeochemistry of ancient and modern environments, Springer-Verlag. Berlin
29. Ghiorse WC, Ehrlich HL (1976) Appl Environ Microbiol 31: 977
30. Francis AJ, Dodge CJ (1988) Appl Environ Microbiol 54: 1009
31. DeVrind JPM, Boogerd FC, DeVrind-DeJong EW (1986) J Bacteriol 167: 30
32. Myers CR, Myers JM (1992) J Bacteriol 174: 3429
33. Bennett JC, Tributsch H (1978) J Bacteriol 134: 310
34. Rodriquez-Leiva M, Tributsch H (1988) Arch Microbiol 149: 401
35. Tributsch H (1976) Naturwissenschaften 63: 88
36. Tributsch H, Bennett JC (1981) J Chem Tech Biotechnol 31: 565
37. Tributsch H, Bennett JC (1981) J Chem Tech Biotechnol 31: 627
38. Ehrlich HL (1993). In: Torma AE, Apel ML, Brierley CL (eds) Biohydrometallurgical technologies. The Minerals, Metals & Materials Society, vol 2, p 415
39. Trimble RB, Ehrlich HL (1987) Appl Microbiol 16: 695
40. Ehrlich HL, Ingledew WJ, Salerno JC (1991). In: Shively JM, Barton LL (eds) Variations in autotrophic life. Academic Press, London, p 147

41. Ingledew WJ, Cobley JC (1980) Biochim Biophys Acta 590: 141
42. Ingledew WJ (1986). In: Ehrlich HL, Holmes DS (eds) Workshop on biotechnology for the mining, metal-refining and fossil fuel processing industries. Biotech Bioeng Symp No. 16, Wiley, New York, p 23
43. Lovley DR, Phillips EJ, Lonergan DJ (1989) Appl Environ Microbiol 55: 700
44. Serebrjanaja M, Yakhontova L, Petrova L (1993). In: Torma AE, Wey JE, Lokshmanan VL (eds) Biohydrometallurgical technologies. The Minerals, Metals & Materials Society. vol 1 p 277
45. Stone AT, Morgan JJ (1984) Environ Sci Technol 18: 617
46. Stone AT (1987) Geochim Cosmochim Acta 51: 919
47. Ehrlich HL (1981) Geomicrobiology, Marcel Dekker, Inc. New York
48. Lovley DR, Phillips EJP (1988) Geomicrobiol J 6: 145
49. Dubinina GA (1979) Microbiology USSR 47: 471
50. Silverman MP, Lundgren DG (1959) J Bacteriol 77: 642
51. Ghosh J, Imai K (1985) J Ferment Technol 63: 295
52. Ghosh J, Imai K (1985) J Ferment Technol 63: 259
53. Troshanov EP (1968) Mikrobiologia 38: 528
54. Noble EG, Baglin EG, Lampshire DL, Eisle JA (1991). In: Smith RW, Misra M (eds) Mineral bioprocessing. The Materials, Metals & Materials Society, p 233
55. Ehrlich HL, Yang SH, Mainwaring Jr. JD (1973) Zeit Allg Mikrobiologie 13: 39
56. Srimekanond A, Madgwick J, Pracejus B (1992) Australs. IMM Proc no 2: 77
57. Mero J (1962) Econ Geol 57: 747
58. Murray JW, Balistrieri LS, Paul B (1984) Geochim Cosmochim Acta 48: 1237
59. Burns RG, Fuerstenau DW (1966) Amer Mineralogist 51: 895
60. Ehrlich HL (1984). In: 2nd International Seminar on the Offshore Mineral Resources. Offshore Prospecting and Mining Problems: Current Status and Future Developments, Germinal, Orleans, France, p 639
61. Rusin PA (1993) Microbial removal of residual metals from mine tailings and dump wastes. National Science Foundation, Phase 1 Final Report. p 16
62. Stone AT, Morgan JJ (1987). In: Stumm W (ed) Aquatic surface chemistry. Wiley-Interscience: New York, p 222
63. Ehrlich HL (1993). In: Bollag JM, Stotzky G (eds) Soil biochemistry. Marcel Dekker, Inc New York, vol 8, p 227
64. Rusin P, Sharp J, Arnold R, Sinclair NA, Young T (1992) Min Eng 44: 1467
65. Hensyl WR (ed) (1994) Bergey's manual of determinative bacteriology. 9th edn. Williams and Wilkens, Baltimore
66. Miller TL, Churchill BW (1986). In: Demain AL, Solomon NA (eds) Manual of industrial microbiology and biotechnology. American Society for Microbiology, Washington, DC, p 122
67. Perkins EC, Novielli F (1962) Bacterial leaching of manganese ores. US Department of the Interior. Bureau of Mines, Report no 6102
68. Kozub JA, Madgwick JC (1983) Proc Aust Inst Min Metall 288: 51
69. Silverio CM, Madgwick JC (1985) Proc Aust Inst Min Metall 290: 63
70. Veglio F, Terreri M, Toro L (1993). In: Torma AE, Wey JE, Lakshmanan VL (eds) Biohydrometallurgical technologies. The Minerals, Metals & Materials Society vol 1, p 269
71. Hart MJ, Madgwick JC (1987) Proc Aust Inst Min Metall 292: 61
72. Srimekanond A, Thangavelu V, Madgwick J (1992) J Ind Microbiol 10: 217
73. Hart MJ, Madgwick JC (1986) Proc Aust Inst Min Metall 291: 61
74. Marshall KC (1979). In: Trudinger PW, Swaine DJ (eds) Biogeochemical cycling of mineral forming elements. Elsevier, New York, p 253
75. Buys H, Chan SM, Chun UH, Cho DW, Davis P, MacKay B, Madgwick JC, Pannowitz D, Poi G, Varga R (1986) Proc Aust Inst Min Metall 291: 71
76. Rusin PA, Quintana L, Brainard JR, Strietelmeier BA, Tait CD, Newton TW, Clark D, Palmer P, Ekberg SA (in press) Environ Sci Technol
77. Collins YE, Stotzky G (1989) Beveridge TJ, Doyle RJ (eds) Metal ions and bacteria, John Wiley & Sons, New York, p 31
78. Holmes DS (1988) Mineral Metall Processing 5: 49
79. Holden PJ, Madgwick JC (1983) Proc Aust Inst Min Metall 286: 61
80. Espejo RT, Romero P (1987) Appl Environ Microbiol 53: 1907
81. Abbruzzese C, Duarte MY, Paponetti B, Toro L (1990) Minerals Eng 3: 307

82. DeHuff GL (1965). In: Mineral facts and problems. Bureau of Mines Bulletin 630. US Depart of the Interior, Washington, DC, p 553
83. Henn JJ, Kirby RC, Norman LD (1968) Review of major proposed processes for recovery of manganese from United States resources. Information circular 8368. Bureau of Mines, US Department of the Interior

Production of Poly(hydroxyalkanoic acid)

Sang Yup Lee* and Ho Nam Chang
Department of Chemical Engineering and BioProcess Engineering Research
Center, Korea Advances Institute of Science and Technology, Daeduk Science
Town, Taejon 305-701, Korea

Poly(hydroxyalkanoic acid) [PHA] is accumulated by numerous microorganisms as an energy reserve material under unbalanced growth conditions in the presence of excess carbon source. In spite of being a good candidate for biodegradable thermoplastics, their high price compared with conventional plastics currently in use has limited their availability in a wide range of applications. With the aim of reducing the high production cost of PHA, much effort is currently being devoted to improve productivity by employing various microorganisms and by developing efficient culture techniques. Several processes recently developed and employed for the production of PHA by various bacteria are described.

---

* To whom correspondence should be addressed

Advances in Biochemical Engineering/
Biotechnology, Vol. 52
Managing Editor: A. Fiechter
© Springer-Verlag Berlin Heidelberg 1995

1 Introduction

Plastic materials that have been universally used in our daily lives are now causing serious environmental problems due to their non-biodegradability. In order to reduce the amount of plastic waste, world-wide programs for efficient management of used-plastic materials such as recycling have been initiated. Another solution to reduce plastic waste is the use of biodegradable plastics. Among the candidates for biodegradable plastics the polymers poly-(hydroxyalkanoic acid) [PHA] have been drawing special attention due to their similar properties to synthetic plastics and excellent biodegradability [1–4]. PHAs are natural storage compounds synthesized and accumulated by numerous microorganisms when cells encounter unbalanced growth condition in the presence of an excessive carbon source [3,5]. Beside poly (3-hydroxybutyric acid) [P(3HB)], a member of PHA that has most often been found, PHA made up of many other 3-, 4- and 5-hydroxyalkanoic acids have been identified as constituents of these polymers in the last several years [6]. PHA can be divided into two groups by the number of carbon atoms in the monomer; short-chain-length PHA consisting of 3–5 carbon atoms (PHA_{SCL}) and medium-chain length PHA consisting of 6–14 carbon atoms (PHA_{MCL}) [6]. The general structural formula of PHA is shown in Fig. 1. The physical and mechanical properties such as brittleness, stiffness, and melting or glass transition temperature were reported to vary considerably among various PHA having different monomer components [4,6]. Even though use of PHA for a wide range of general, agricultural, marine and medical applications have been suggested [2,7,8], actual application has been limited by their high price compared with conventional plastic materials. Current research on PHA can be divided into several areas (for reviews see: 4–6, 9–11]): development of PHA having novel constituents including the recently found 4-hydroxyhexanoic acid [12]; characterization of

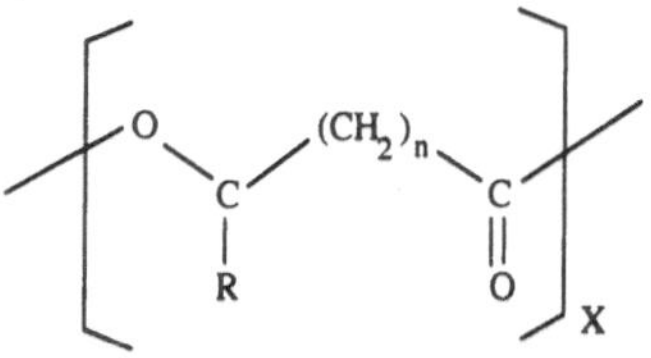

X=100 - 30000

n=1, R=CH₃   --> P(3HB)
    R=-CH₂-CH₃   --> P(3HV)
    R=-(CH₂)₄-CH₃   --> P(3HO)
n=2, R=H   --> P(4HB)
    R=-CH₃   --> P(4HV)
n=3, R=H   --> P(5HV)

Fig. 1. General structural formula of poly(hydroxyalkanoic acid) [3,6]

PHA polymer in vivo and in vitro; physiology, genetics and molecular biology of microorganisms synthesizing PHA; production of PHA by high cell density culture; isolation and purification of PHA. The latter two are directly related to possible cost reduction. To date, more than 250 different microorganisms are known to synthesize and accumulate various PHA [6]. However, only a few bacteria having the ability of accumulate PHA to a significant level have been employed for the production of PHA. Among these are *Alcaligenes eutrophus* [1], *Alcaligenes latus* [13], *Azotobacter vinelandii* [14], a couple of methylotrophs [15–17], *Pseudomonas oleovorans* [18], and recombinant *Escherichia coli* [19, 20]. Different strategies for growing these bacteria to high density, which is necessary for obtaining high concentration of PHA, have been developed, and conditions required for efficient PHA accumulation such as nutrient limitation have been identified for some of them. Several processes developed for the production of various PHAs employing these bacteria are reviewed.

2 Production Strategies

Since PHA are synthesized and accumulated under unfavorable growth condition in the presence of excess carbon source [5], it is important to develop cultivation strategies that simulate these conditions and allow efficient production of PHA. There are numerous microorganisms that accumulate PHA, and obviously, the optimal condition for the production of PHA will be different from one organism to another. Some microorganisms accumulate PHA under nutrient limitation, while some others such as *A. latus* accumulate polymer during growth [13]. It is also important to consider the cyclic nature of PHA metabolism [5, 21], synthesis and degradation, since unnecessarily prolonged cultivation may result in the degradation of PHA by depolymerases [22].

Selection of an organism for the industrial production of PHA will be based on a number of factors such as the cell's ability to utilize cheap substrate, growth rate, polymer synthesis rate and the extent of polymer accumulation. The latter three will directly affect the productivity, which should be as high as possible. The ability to use cheap carbon sources can significantly reduce the cost of PHA. The yield of PHA on carbon source is also important so as not to waste substrate to non-PHA cellular materials. Several cheap carbon sources are compared for their prices and the approximate P(3HB) yields on these substrates (Table 1, [1]). Basic strategies for PHA production can be developed based on the consideration of yield and productivity as recently reviewed [23, 24]. Another thing that should be considered is the recovery of PHA, which significantly affects the overall economics. Some PHA-producing organisms also produce substantial amount of other polymers such as polysaccharide [25]. These organisms should be avoided since they not only waste carbon sources but also make the recovery process inefficient. After consideration of these factors, the

Table 1. Substrate prices and the yields of P(3HB) [1]

Substrate	Apprpximate price (£ per t)	Yield (P(3HB) g g^{-1} substrate)
Acetic acid	370	0.33
Ethanol	440	0.5
Glucose	200	0.33
Hydrogen	500	1.0
Methanol	90	0.18
Sucrose	200	0.33

organisms described below seem to be suitable candidates for the production of PHA. Cultivation strategies developed for these organisms are now described.

3 *Alcaligenes eutrophus*

Since the first discovery of P(3HB) by Lemoigne in 1926, numerous bacteria have been identified which accumulate various PHAs [6]. *Alcaligenes eutrophus* has been studied in most detail due to its ability to accumulate large amount of P(3HB) (ca. 80%, wt/wt of dry cell mass). It is ironical to find that early studies on *A. eutrophus* were focused on eliminating P(3HB) rather than overproducing it, for the production of single cell proteins. The PHA metabolism of this chemolithoautotropic aerobic bacterium has been well understood. Three enzymes are involved in the conversion of acetyl-CoA to P(3HB): β-ketothiolase (EC 2.3.1.9), NADPH dependent acetoacetyl-CoA reductase (EC 1.1.1.36) and PHA synthase [5,6]. The *A. eutrophus* PHA biosynthetic pathway is shown in Fig. 2. In 1981, Imperial Chemical Industries (ICI, UK) found that a copolyester

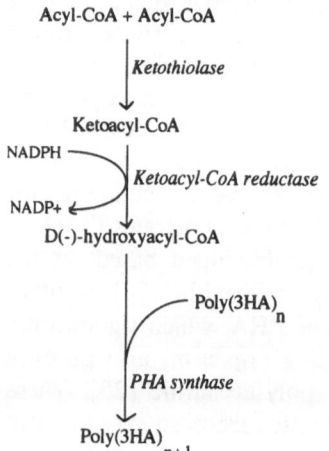

Fig. 2. *Alcaligenes eutrophus* PHA biosynthetic pathway

of 3-hydroxybutyric acid and 3-hydroxyvaleric acid, P(3HB-*co*-3HV), could be synthesized by *A. eutrophus* from glucose and propionic acid [1, 26]. Since then, a variety of copolymers have been synthesized by *A. eutrophus* from various carbon sources [3, 27].

3.1 Industrial Production

After the discovery of the thermoplastic properties of P(3HB), W.R. Grace and Co. in the USA produced P(3HB) for the possible commercial applications in the early 1960s [28, 29]. However, the company shut down the process due to the low production efficiency and the lack of suitable recovery methods, and industrial interest on PHA lay dormant until ICI initiated the project in the 1970s. ICI published a number of patents on the production and recovery of PHA, and is currently producing P(3HB) and P(3HB-*co*-3HV) from glucose and a mixture of glucose-propionic acid, respectively, by fed-batch culture of a glucose utilizing mutant strain of *A. eutrophus* H16 on a fairly large scale [12, 30, 31]. A plant that produces 300 tonnes of P(3HB-*co*-3HV) annually is in operation, and a new one capable of producing 10 000 t annually is planned [32].

ICI produces PHA by the strategy of nutrient limitation, and the process can be divided into growth and polymer accumulation phases. A number of nutritional components such as nitrogen and phosphate were examined for the effect of their limitation on the synthesis and accumulation of PHA, and phosphate was chosen as a limiting factor in ICI process [1]. *A. eutrophus* is grown in a glucose-salts medium containing phosphate, the amount of which is limited to support only a desired amount of cell growth. After about 60 h of cell growth, phosphate limitation is reached. During the next 40–60 h, glucose is added to accumulate P(3HB). Cultures have been carried out in both airlift and stirred reactors (up to 220 m^3). Total dry cell weights (which contains P(3HB)) of more than $100 \, \mathrm{g \, l^{-1}}$ having the P(3HB) content of 70–80% are routinely obtained [1, 31]. Operations for the production of P(3HB-*co*-3HV) are similar to that described above except that both glucose and propionic acid are fed in the polymer accumulation phase. The content of 3-hydroxyvaleric acid could be controlled by variation of the ratio of glucose to propionic acid in the feed. Normally, P(3HB-*co*-3HV) copolymers containing 0–30 mol% of 3-hydroxyvaleric acid are produced [1, 31]. Copolymers having higher fraction (up to 90 mol %) of 3-hydroxyvaleric acid have been produced from butyric and/or pentanoic acids [27], but production in an industrial scale has not been reported.

Ethanol has also been considered as a substrate for the production of P(3HB) by a mutant strain of *A. eutrophus* [1, 33]. P(3HB-*co*-3HV) could be produced by feeding propanol along with ethanol under nitrogen limiting conditions [34]. Recently, there has been a report on the production of P(3HB) and P(3HB-*co*-3HV) by fed-batch culture of *A. eutrophus* NCIMB 12080 from alcohols. The final concentration of P(3HB) and polymer content obtained in 50 h were $47 \, \mathrm{g \, l^{-1}}$ and 74%, respectively [35]. In an industrial point of view, ethanol is not a good substrate for the production of PHA due to its higher price than carbohydrates (see Table 1).

3.2 Poly (3-hydroxybutyric acid) from Gases

A. eutrophus H16, which is the parent strain of the glucose- or alcohol-utilizing mutant strains employed by ICI, can grow on gas mixtures of hydrogen/carbon dioxide/air [5,36,37]. The use of hydrogen/carbon dioxide as a substrate has been considered to be an economically attractive process. However, an investigation of the engineering problems associated with plant construction and operation for the safe use of the flammable gas mixture led to the conclusion that this process is uneconomical due to the high capital costs [1]. Nevertheless, a group of scientists investigated the production of P(3HB) from hydrogen/carbon dioxide. The recycled gas closed circuit culture system was developed, which allowed complete consumption of the gases supplied [38]. It was found that oxygen limitation was superior to ammonium limitation for the production of P(3HB) [39]. Cell mass and P(3HB) concentration of 85 and $61.5 \, g \, l^{-1}$, respectively, were obtained from hydrogen/carbon dioxide mixture after 40 h of cultivation [40]. Even though oxygen concentration in the culture system was maintained below the lower limit for detonation, there was always a chance of explosion. To solve this dangerous problem, they developed a two stage culture method, which separates the growth and the polymer accumulation phases. Only a limited amount of oxygen is supplied during the polymer accumulation phase. The PHB concentration and productivity of $20 \, g \, l^{-1}$ and $0.9 \, g \, l^{-1} \, h^{-1}$ was obtained [40]. This system is currently under investigation for improved continuous operation.

3.3 Strategies for High Cell Density Culture

The most popular method for achieving a high cell density, which is often necessary for high productivity of the desired product, has been fed-batch culture [41]. Continuous culture systems may offer higher productivity than fed batch cultures, but only when the culture can be maintained stably without contamination. Membrane bioreactors have also been employed to obtain a high cell density but are still awaiting further improvement to be used on an industrial scale [42]. During fed-batch culture, nutrients are usually fed into the reactor intermittently or constantly by monitoring one of several feedback parameters such as dissolved oxygen (DO) or pH [41]. In these approaches, however, substrate concentration in the reactor cannot be precisely maintained due to the nature of indirect estimation. From the kinetic studies for the growth of A. eutrophus and PHA accumulation by us and others [43], it was found that the concentration of the carbon source should be maintained at the optimum value for efficient production of PHA. Therefore, the development of a method to monitor precisely and control the concentrations of the carbon source is essential. The two most popular methods to monitor the substrate concentration precisely are estimation by carbon dioxide evolution rate (CER) and an automatic on-line monitoring system [44–46]. Suzuki et al. [15] developed a method to control methanol concentration for the production of P(3HB) by a methylotroph as will be seen later.

The strategies we have taken to produce P(3HB) and P(3HB-*co*-3HV) to high concentrations by fed-batch culture of *A. eutrophus* NCIMB 11599, a glucose-utilizing mutant of *A. eutrophus* H16, are described below. Two methods were used to maintain the glucose concentration at the optimum value: one was the use of CER measured by mass spectrometry to estimate glucose concentration, and the other was the use of an on-line glucose analyzer to directly control glucose concentration in the medium [47–49]. The schematic diagram of the system employing an on-line glucose analyzer is shown in Fig. 3. Glucose concentration was maintained at $10–20 \, gl^{-1}$, which was optimal for the growth of this strain. The strategy employed for triggering polymer accumulation was nitrogen limitation. The ammonia solution was used to control pH and to supply nitrogen source during the cell growth phase. When cell mass reached a desired value, the ammonia solution was replaced by a mixture of NaOH/KOH solution to apply nitrogen limitation. It was thought to be important to find when to apply nitrogen limitation, since early application would result in a low final concentration of PHA and late application might cause incomplete accumulation of PHA with a large amount of non-PHA cellular materials. Therefore, a series of fed-batch cultures were carried out for the production of P(3HB), where nitrogen limitation was applied at different cell densities. It was found that the final P(3HB) concentration and productivity increased as nitrogen limitation was applied at a higher cell concentration up to $70 \, gl^{-1}$. Further delaying nitrogen limitation until the cell concentration reached $90 \, gl^{-1}$ resulted in lower polymer concentration due to culture instability. The results

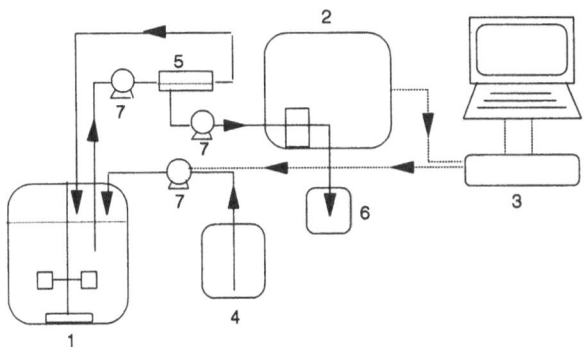

Fig. 3. Schematic diagram of the culture system with on-line glucose analyzer: *1*, Reactor; *2*, glucose analyzer; *3*, personal computer; *4*, feed reservoir; *5*, membrane module; *6*, waste vessel; *7*, peristaltic pump

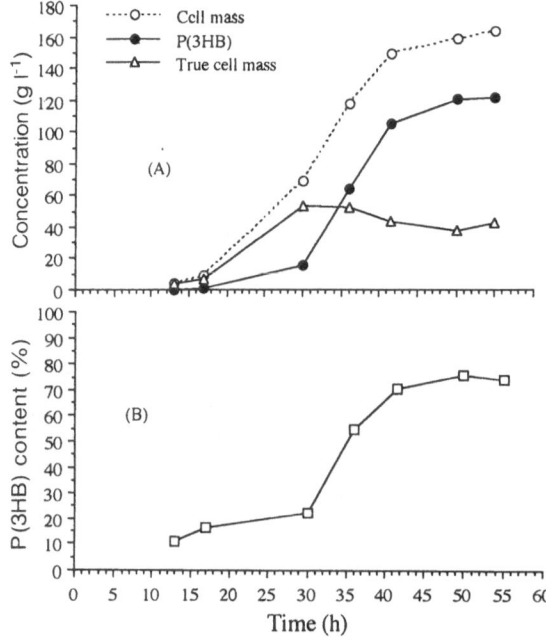

Fig. 4. A,B Production of P(3HB) by fed-batch culture of *A. eutrophus* NCIMB 11599 with glucose concentration control: **A)** time profiles of cell mass, P(3HB), and true cell mass; **B)** time profile of P(3HB) content [48]

obtained by applying nitrogen limitation at the cell mass of 70 g l^{-1} are shown in Fig. 4. The final cell mass, P(3HB) concentration, and PHB content obtained in 50 h were 164 g l^{-1}, 121 g l^{-1}, and 76%, respectively [48]. This resulted in the productivity of 2.42 g l^{-1} h^{-1}, which is the highest value among those reported.

Fed-batch cultures of *A. eutrophus* for the production of P(3HB-*co*-3HV) were also carried out using similar feeding strategy described above. Cells were grown to a concentration of 60–70 g l^{-1} by feeding glucose only using an on-line glucose analyzer. During the polymer accumulation phase the feeding solution was changed to a mixture of glucose and propionic acid having different P/G ratios (the ratio of propionic acid to glucose in the feeding solution). The results obtained with the feed having the P/G ratio of 0.17 are shown in Fig. 5. The final cell mass and PHA concentration decreased from 158 and 117 to 113 and 64 g l^{-1}, respectively, as the P/G ratio increased from 0.17 to 0.52 (mol mol^{-1}). The PHA content in the cell and the PHA productivity also decreased from 74 to 56.5% of cell mass and from 2.55 to 1.64 g l^{-1} h^{-1}, respectively [49]. The fraction of 3-hydroxyvaleric acid in the copolymer, however, continuously increased to 14.3 mol% with increasing P/G ratio. Propionic acid concentration in the medium was maintained below 1.3 g l^{-1} when the P/G ratio was 0.17 or 0.35 (mol mol^{-1}). The lower polymer content with the higher P/G ratio suggests that propionic acid inhibited polymer synthesis. The yields of 3-hydroxyvaleric acid from propionic acid and total PHA from two substrates were calculated from the amounts of glucose and propionic acid fed into the bioreactor. Both yields decreased with increasing propionic acid feeding rate. With the P/G ratios

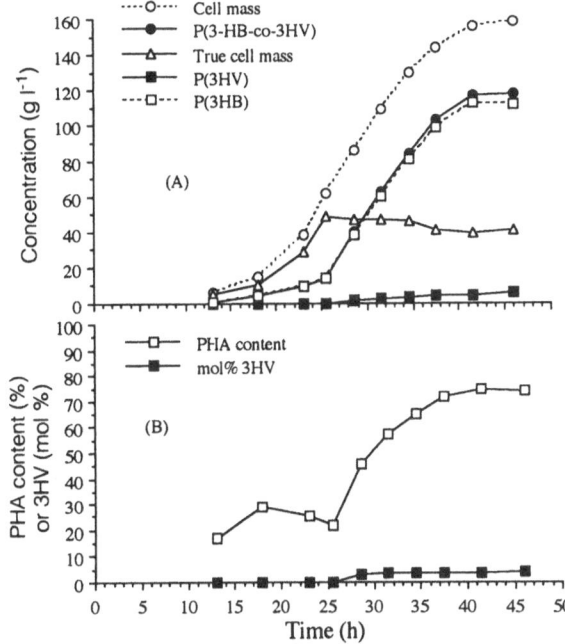

Fig. 5. Production of P(3HB-co-3HV) by fed-batch culture of A. eutrophus NCIMB 11599 by feeding glucose and propionic acid (P/G ratio of 0.17) with glucose concentration control: A) time profiles of cell mass, P(3HB-co-3HV), true cell mass, P(3HB), and P(3HV); B) time profiles of PHA content and mole % 3HV [49]

of 0.17, 0.35, and 0.52, the final yields of 3-hydroxyvaleric acid from propionic acid were 0.344, 0.324, and 0.293 (g 3-hydroxyvaleric acid per g propionic acid), respectively. The yields of total PHA from two substrates were 0.3, 0.221, and 0.2 (g PHA per g substrates), respectively [49].

4 Alcaligenes latus

In the mid 1980s an Austrian company Chemie Linz AG also started a project for the production of P(3HB). They employed a newly isolated strain of A. latus, which was found to accumulate P(3HB) during the growth phase [13, 50]. The polymer team of Chemie Linz, which later became Petrochemie Danubia, published several patents on the production of P(3HB) by A. latus [51–54]. The biotechnological research unit, Biotechnologische Forschungsgesellschaft mbH (btF), affiliated with Chemie Linz and other Austrian companies developed a process for producing one tonne of P(3HB) a week in a 15 m^3 reactor [50]. The strain used is A. latus btF-96, which is a mutant strain possessing higher P(3HB) productivity derived from A. latus DSM1124 [13, 50]. Some advantages of employing A. latus compared with A. eutrophus are faster growth, ability to use cheap sucrose (beet and cane molasses) as a carbon source and to accumulate P(3HB) during growth. Production of more than 60 g l^{-1} of P(3HB) by fed-batch culture of A. latus btF-96 from sucrose has been reported [50]. The

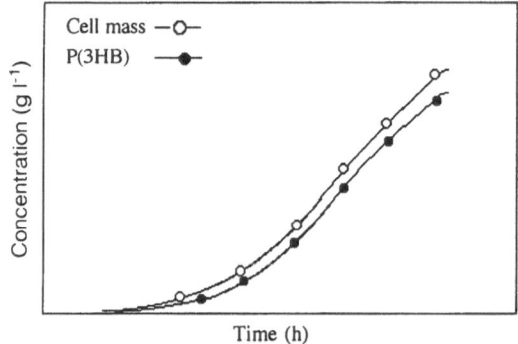

Fig. 6. P(3HB) accumulation during cell growth in fed batch culture of *A. latus* [13]

patterns of cell growth and P(3HB) accumulation are shown in Fig. 6. Recently, professor Yamane's group developed a continuous culture system that allowed production of P(3HB) with very high productivity of 2.6 g l^{-1} h^{-1} using *A. latus* DSM1123, one of the patented strain of Petrochemie Danubia (Personal communication of Yamane T, Nagoya Univ., Japan). Reports on the strain stability during the prolonged cultivation are to be seen. If cell growth and polymer accumulation are stable for a long period of time, this process may be a good alternative to the one currently in use.

5 *Azotobacter vinelandii*

In the late 1960s, *Azotobacter* species were found to accumulate large amounts of PHA [55]. Oxygen limitation, rather than the more frequently encountered nitrogen limitation, allowed accumulation of PHA in *Azotobacter beijerinckii* [56,57]. Studies on the regulation of P(3HB) metabolism in *A. beijerinckii* revealed than P(3HB) synthesis and degradation were controlled by the cellular redox state and the availability of CoASH, pyruvate and α-oxoglutarate [58]. The first enzyme in the three-step PHA biosynthesis pathway, β-ketothiolase, was found to be important in the regulation of PHA metabolism in *Azotobacter*.

The group of W. Page has been working on the production of PHA by employing *A. vinelandii*. The strain used in their studies is a mutant strain *A. vinelandii* UWD (ATCC 53799), which is able to accumulate large amount of P(3HB), up to 75% of dry cell mass, during exponential growth [59]. Polymer synthesis in this mutant is not dependent on oxygen limitation, but the efficiency of converting glucose to P(3HB) is enhanced in oxygen-limiting condition [59]. The yield of P(3HB) obtained by *A. vinelandii* UWD was 0.33 g P(3HB) per g glucose, which can be compared with the yield of 0.05 obtained by wild type strain UW [59]. This strain was patented [60], and has been used in the subsequent polymer production studies by the same group.

A. *vinelandii* UWD was cultured in the media containing various complex carbon and nitrogen sources for the production of P(3HB) [61, 62]. It was found that unrefined carbon sources such as beet and malt molasses supported good formation of P(3HB), and the yields of P(3HB) were higher than those obtained with refined carbon sources [61]. Beet molasses, which is about one-third of the price of glucose, is attractive since it promoted polymer synthesis and allowed P(3HB) accumulation up to 7 g l^{-1} in flask culture [14]. Copolymer P(3HB-*co*-3HV) could also be synthesized in *A. vinelandii* by feeding valeric or heptanoic acid during polymer synthesis in beet molasses medium [63]. By varying the concentration of valeric acid in the feed, copolymers having as high as 23 mol% hydroxyvaleric acid units could be obtained [63]. Increased yields of P(3HB) could be obtained in the media containing pure or unrefined carbon sources by the addition of complex nitrogen sources [62]. Addition of 0.05 – 0.2% fish peptone, proteose peptone or yeast extract resulted in the greatest enhancement of P(3HB) synthesis and accumulation (up to 25 fold increase) among the tested [62]. The PHA promoting effect of complex nitrogen sources was most significant in well-aerated cultures, and could be explained by reduced cellular demands for some de novo amino-N synthesis upon the addition of these nitrogen sources [14]. The effects of supplementing complex nitrogen sources (2 g l^{-1}) on P(3HB) synthesis by *A. vinelandii* in glucose medium are summarized in Table 2. It is interesting to note that these nitrogen sources enhanced PHA synthesis without general growth promotion. Fish peptone, which allowed the highest enhancement of PHA synthesis, was employed in fed-batch cultures [64]. Continuous supplementation of fish peptone along with glucose allowed enhanced P(3HB) production with high yield up to 0.65 g P(3HB) per g glucose [64]. The best conditions for P(3HB) production were obtained by starting the culture without fish peptone and then feeding the mixture of glucose and fish peptone. The final P(3HB) concentration of 32 g l^{-1} could be obtained in 47 h with relatively high yield of 0.34 g P(3HB) per g glucose [64]. However, the productivity was only 0.68 g P(3HB) $l^{-1} h^{-1}$, which is rather low compared with that obtained by employing *A. eutrophus* or *A. latus*.

A. *vinelandii* UWD cells cultivated in the medium containing fish peptone became fragile, and P(3HB) could be simply extracted by treatment with

Table 2. Effect of complex nitrogen sources on P(3HB) formation by *A. vinelandii* [62]

Complex N Source	P(3HB) (g l^{-1})	Residual Biomass (g l^{-1})	P(3HB) Content (%)
None	0.3	1.6	16
Fish peptone	7.5	2.6	74
Yeast extrast	6.8	2.5	73
Tryptone	5.5	2.1	72
Casamino acids	4.7	2.2	68
Bacto peptone	4.7	2.2	68
Beef extract	3.8	1.9	67

1 N aqueous NH_3 (pH 11.4) at 45 °C for 10 min [64]. The polymer mixture obtained by this treatment contained 94% P(3HB), 2% protein, and 4% residual cell material [64], which proves to be an efficient purification method. In our opinion *A. vinelandii* UWD is a good candidate for the production of PHA due to its ability to use cheap substrates and efficient purification process, but the PHA productivity should be increased to a comparable value obtained with *A. eutrophus* or *A. latus* by developing better cultivation strategy.

6 Methylotrophs

Methanol is one of the cheapest carbon substrates that can be utilized in production processes (See Table 1). ICI, during the early studies, considered use of methylorophs for the production of P(3HB) from methanol because the company had experience in methanol process technology [1]. The process employing several strains of *Methylobacterium organophilum* for the production of P(3HB) has been patented [65]. However, relatively low amount of polymer accumulation by methylotrophs made the recovery process difficult since more biomass had to be processed per tonne of P(3HB) [1]. Nevertheless, the low price of methanol has attracted a number of researchers to devise a process for the production of PHA by employing newly isolated strains of methylotrophs and by developing efficient cultivation techniques.

The group led by T. Yamane reported impressive results on the production of P(3HB) from methanol [15, 66]. They were able to produce very high concentration of P(3HB), 136 g l^{-1}, by fully automatic fed batch cultures of *Protomonas extorquens* K using methanol as a carbon source [66]. During the cultivation, temperature, dissolved oxygen concentration, and methanol concentration were automatically controlled at 30 °C, 2.5 ppm, and 0.5 g l^{-1}, respectively, and other nutrients including nitrogen source were also controlled at their initial concentrations. The effect of nitrogen source (ammonia water) on the production of P(3HB) was examined. It was found that a nitrogen source was required even in polymer accumulation phase [67], which is different from *A. eutrophus*. By controlling the C/N (carbon to nitrogen) ratio during the fed batch culture, the final cell and P(3HB) concentrations of 233 and 149 g l^{-1}, respectively, could be obtained in 170 h with the yield of 0.2 g P(3HB) per g methanol [15]. However, the P(3HB) productivity was 0.88 g P(3HB) l^{-1} h^{-1}, which is lower than that obtained with *A. eutrophus* or *A. latus*. The average molecular weight of P(3HB) obtained by fed-batch culture of *P. extroquens* varied significantly from 5×10^4 to 8×10^5 with changing methanol concentration [68]. Higher molecular weight P(3HB) was obtained with lower methanol concentration in the medium [68]. This bacterium was later named as *Methylobacterium extorquens* K [16]. *M. extorquens* K and another methylotrophic bacterium *Paracoccus denitrificans* were shown to synthesize the copolymer P(3HB-*co*-3HV) when methanol and *n*-amyl alcohol were used together as carbon sources in nitrogen-limited medium [16, 69]. The hydroxyvaleric acid content in the copolymer varied considerably between the two strains: it increased with increasing concentration of *n*-amyl alcohol in *P. denitrificans* leading to the maximum content of 91.5 mol%, while it was limited to the maximum of 38.2 mol% in *M. extorquens* [16]. The genes involved in the PHA biosynthesis were recently cloned from *P. denitrificans*, and are expected to reveal the characteristics of PHA biosynthesis enzymes at molecular level (Personal communication of Yamane T, Nagoya Univ., Japan). Even though not reported yet, it seems possible to obtain high concentration of the copolymer from methanol and *n*-amyl alcohol by fed batch culture with controlled feeding of both substrates as done previously.

There have been several other groups who have studied PHA production by different methylotrophs. *Pseudomonas* 135 was able to accumulate P(3HB) up to 55% in nitrogen limited fed batch culture [17]. *Methylobacterium sp.* (KCTC 0048) was shown to synthesize various copolyesters when secondary carbon sources such as valeric acid, pentanol, 4-hydroxybutyric acid, or 1,4-butanediol were added to nitrogen-limited cultures containing methanol as a major carbon source [70]. Fed batch cultures of these bacteria have not been reported. A Canadian group also reported production of P(3HB) and P(3HB-*co*-3HV) by employing a new isolate of *M. extorquens* [71]. Cell and P(3HB) concentrations of 120 and 60 g l^{-1}, respectively, were obtained by fed batch culture with methanol concentration control at 1.4 g l^{-1}.

Use of methylotrophs for the production of PHA seems to be attractive since high concentration of PHA can be obtained from the cheap substrate methanol. However, it should be remembered that the PHA content obtained with methylotrophs was generally lower than that obtained with *Alcaligenes* or *Azotobacter*. Furthermore, the highest PHA productivity obtained with *M. extroquens* K (0.88 g l^{-1} h^{-1} PHA) [15] is still much lower than that obtained` with *A. eutrophus* (2.42 g l^{-1} h^{-1} PHA) [48]. These need to be improved for the production of PHA at low cost, and to fully appreciate the use of the cheap substrate methanol. Also, one should remember that the average molecular weights of PHA synthesized by methylotrophs (in the order of 10^5) are generally lower than those obtained with *Alcaligenes* or *Azotobacter*.

7 Pseudomonads

As briefly mentioned in the introduction, various PHA having 3-, 4-, and 5-hydroxyalkanoic acids have been identified as constituents of the bacterial polyesters [5,6]. Pseudomonads are widespread in many natural environments, and have been shown to be capable of synthesizing medium-chain-length PHA (PHA$_{MCL}$) consisting of 6–14 carbon atoms [72,73]. Advances in the production of PHA$_{MCL}$ by pseudomonads are summarized in this section.

7.1 Medium-Chain-Length Poly(hydroxyalkanoic acid)

PHA having various medium-chain-length 3-hydroxyalkanoic acids were first detected in cells of *Pseudomonas oleovorans* ATCC 29347 grown on *n*-octane [74]. PHA containing as many as six different types of monomer units having 6–11 carbon atoms were formed in *P. oleovorans* grown on C_6–C_{10} *n*-alkanoic acids [75]. The maximum polymer content was 30% and the average molecular weights ranged from 9×10^4 to 3.7×10^5 [75]. When *P. oleovorans* was grown on C_6 to C_{12} *n*-alkanes and 1-alkenes, PHA_{MCL} containing saturated and saturated plus unsaturated monomers, respectively, were synthesized [76]. The fraction of unsaturated monomer units could be modulated by varying the ratio of alkenes to alkanes in the medium [76], which may be useful for the development of various polymers that can be chemically modified. The search for other pseudomonads accumulating PHA has revealed that all of the fluorescent pseudomonads belonging to rRNA homology group I were able to synthesize PHA_{MCL} [72,73,77]. The composition of these PHA_{MCL} was dependent on the length of the substrate carbon backbones [72, 77]. PHA_{MCL} is also synthesized by most of the fluorescent pseudomonads except *P. oleovorands* from unrelated carbon substrates such as carbohydrates [77,78]. It was suggested that the monomers of PHA_{MCL} were derived from the fatty acid β-oxidation pathway when C_6–C_{10} alkane, alkanol, or alkanoic acid was used as substrates, and from the de novo fatty acid synthesis pathway when carbohydrates or gluconate was used [75–78]. PHA_{MCL} containing saturated and unsaturated monomers synthesized by *P. putita* KT2442 grown on glucose indirectly supported the latter hypothesis [79]. The above hypotheses were found to be correct from the results of ^{13}C NMR studies of *P. putita* fatty acid metabolic routes using radiolabeled carbon substrates, 1-^{13}C-decanoic and 1-^{13}C-acetic acid [80]. Using specific inhibitors of fatty acid metabolic pathway, the β-oxidation and de novo fatty acid synthesis were found to function independently in PHA synthesis. Using 1-^{13}C-hexanoic acid as a substrate, it was found that both fatty acid metabolic pathways were operational in PHA synthesis [80].

An interesting observation was recently reported with *Pseudomonas* strain GP4BH1, which was able to utilize citronellol [81]. The strain GP4BH1 accumulated a polyester consisting of 3-hydroxybutyric and medium-chain-length 3-hydroxyalkanoic acids from various carbon sources. The detection of an organism that synthesizes PHA containing both short and medium-chain-length monomer units is notable, since it has been suggested that PHA_{SCL} and PHA_{MCL} are exclusive each other probably due to the substrate specificity of PHA synthases [6]. It is more interesting to see that the polyester accumulated was a blend rather than a copolyester. It was later found that this strain contained two different PHA synthases with different substrate specificity [82].

Another fluorescent pseudomonad, *P. resinovorans* ATCC 14235, was able to synthesize PHA containing C_4–C_6–C_8–C_{10} 3-hydroxyalkanoic acid monomers from hexanoic or octanoic acid, but not from glucose [83]. *P. aeruginosa* AO-232 accumulated PHA containing C_6–C_8–C_{10}–C_{12} monomer units from acetic acid, and PHA composed of 7 different units of C_6–C_{12} from propionic acid [84]. Synthesis of various PHA consisting of unusual monomers such as bromoalkanoic or cyanoalkanoic acid have also been reported (for reviews see [85] and references therein). Two other pseudomonads showing interesting properties are briefly described below. *P. acidovorans* accumulated the polyester P(3HB-*co*-4HB-*co*-3HV) from 1,4-butanediol and pentanol. The fraction of 4-hydroxybutyric acid could reach as high as 99 mol % when 1,4-butanediol was used as the sole carbon source [86]. *P. pseudoflava* was able to produce PHA from pentose sugars, which provides the potential of producing PHAs from cheap substrates such as lignocellulosic biomass [87]. The PHA_{MCL}, represented by poly (3-hydroxyoctanoic acid), has drastically different mechanical properties compared with P(3HB) or P(3HB-*co*-3HV), and can be considered as a unique thermoplastic elastomer. These properties can be found in [88].

7.2. High Cell Density Culture of Pseudomonads

Production of PHA_{MCL} by high cell density culture of *P. oleovorans* has been studied by the group of B. Witholt, who has had much experience with this organism and published a patent on a process for producing polyesters by culturing this organism [89]. They carried out continuous cultures of *P. oleovorans* ATCC 29347 in two-liquid phase media containing 15% (v/v) *n*-octane. Cells were grown in an ammonium-limited condition at varying dilution rates, and cell concentration and PHA content were determined as a function of growth rate [90]. Cell concentration decreased from 2.25 to 1.32 g l^{-1}, while the PHA content decreased from 46.7 to 8.3% of dry cell weight when the dilution rate increased from 0.09 to 0.46 h^{-1} [90]. To increase the steady-state cell concentration, and subsequently the PHA productivity, the concentration of the limiting nutrient ammonium was increased from 16.7 to 116.7 mM. With the optimization of medium and an increase of oxygen transfer rate, the steady-state cell concentration could be raised to 11.6 g l^{-1}, with the PHA productivity of 0.58 g l^{-1} h^{-1} [91].

High cell density cultures of *P. oleovorans* in fed-batch mode were also carried out in the medium containing octane as a carbon source [18]. To obtain high cell densities, it was required to design a reactor allowing very efficient oxygen transfer. The volumetric oxygen transfer coefficient (k_La) was as high as 0.49 s^{-1} in their reactor system at the air flow rate and stirred speed of 2 l min^{-1} and 2500 rpm, respectively [18]. Using this reactor system with optimal feeding of magnesium and the limiting nutrient ammonium, cell and PHA concentrations reached in 38 h were 37.1 and 12.1 g l^{-1}, respectively. The PHA content and PHA productivity were 33% and 0.32 g l^{-1} h^{-1} PHA, respectively [18]. The time courses of cell mass and PHA concentrations are shown in Fig. 7. Further increase of cell and PHA concentrations was hampered by oxygen limitation.

We also carried out fed-batch cultures of *P. oleocorans* using octanol and octanoic acid as carbon sources. Due to the limitation of oxygen transfer

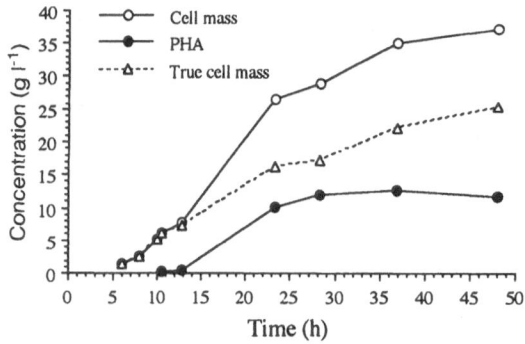

Fig. 7. Time profiles of cell mass, PHA concentration, and true cell mass during fed batch culture of *P. oleovorans* in a two-phase media containing *n*-octane (redrawn from [18])

capacity in our reactor system, pure oxygen was used to meet the needs of cellular oxygen demand. This is why we did not use octane as a carbon source since it is dangerous to use pure oxygen together with this flammable solvent. During the fed-batch culture nutrients consisting of octanoic acid, ammonium sulfate and magnesium sulfate were added intermittently. Cell mass, PHA concentration, and PHA content obtained in 45 h were 41.8 g l^{-1}, 15.5 g l^{-1}, and 37.1%, respectively (unpublished results). Cell mass did not increase further in our system due to the accumulation of octanoic acid, which has been shown to inhibit cell growth at the concentration of 4.65 g l^{-1} (sodium octanoacte) [92]. With octanol as a carbon source, which is much cheaper than octanoic acid, PHA concentration of ca. 14 g l^{-1} could be obtained in 40 h (unpublished results). Octanol may be considered to be a good substrate for the production of PHA since it is much safer to use than octane and its price is reasonable. However, we found one problem in using octanol during the purification of PHA after cultivation. It was difficult to separate cells from the medium even by high-speed centrifugation. A large fraction of cells were trapped in the octanol layer after centrifugation.

It can be seen that much improvement is required for the economical production of PHA by *P. oleovorans*. Continuous culture often results in higher productivity than fed-batch culture, providing that the strain to be cultured is stable during the prolonged cultivation. Since *P. oleovorans* was stable in continuous culture for at least 1 months [91], continuous cultivation may be a better choice for *P. oleovorans* if it turns out to be difficult to further increase cell mass by fed-batch culture.

8 Recombinant *Escherichia coli*

There have been numerous reports of employing recombinant *E. coli* for the production of various bioproducts, mainly proteins (for reviews see [93,94]), since *E. coli* has been the most studied microorganism in every aspect. Recent advances in the production of PHA by recombinant *E. coli* are reviewed in this section.

8.1 Host-Plasmid System

The PHA biosynthesis genes of *A. eutrophus* were cloned in *E. coli* by three independent groups in 1988–89 [95–98]. These genes were sequenced and characterized in detail, and were found to form an operon (Fig. 8, [98,99]), which could be expressed from its own promoter in *E. coli* [99, 100]. Use of recombinant *E. coli* harboring the PHA biosynthesis genes is attractive for the production of PHA since a P(3HB) accumulation of ca. 90% of dry cell weight was achieved in flask culture [100].

We have been carrying out production of PHA by high cell density culture of recombinant *E. coli* harboring the *A. eutrophus* PHA biosynthesis genes [20, 101–105]. To investigate the effects of plasmid copy number (gene dosage) and plasmid stability of PHA accumulation in recombinant *E. coli*, several plasmids having different properties were constructed. Two plasmids, pSYL101 and pSYL102, were constructed by cloning the *A. eutrophus* PHA operon into the high copy number (100–500 copies per cell) plasmid pGEM-7Zf(+) (Promega, WI) and the medium copy number (10–50 copies per cell) plasmid pBR322, respectively [20]. The *E. coli* strain XL1-Blue (Stratagene, CA) harboring pSYL101 and pSYL102 were compared for their ability to accumulate P(3HB) in LB containing $20\,\mathrm{g\,l^{-1}}$ glucose and $0.1\,\mathrm{g\,l^{-1}}$ ampicillin. After 48 h of flask culture, the final cell mass, P(3HB) concentration, and P(3HB) content obtained with XL1-Blue (pSYL101) were $6.2\,\mathrm{g\,l^{-1}}$, $4.6\,\mathrm{g\,l^{-1}}$, and 74.2%, respectively, while those obtained with XL-1-Blue (pSYL102) were $4.9\,\mathrm{g\,l^{-1}}$, $2.4\,\mathrm{g\,l^{-1}}$, and 49%, respectively [20]. The copy numbers of pSYL101 and pSYL102 were 280 and 21 copies per cell, respectively, when the optical density (at 600 nm) of the medium was 3. Therefore, high gene dosage was requires for the production of P(3HB) to a high concentration in recombinant *E. coli*. This was supported by others who showed that higher concentration (up to 40 times) of P(3HB) was obtained using a high copy number plasmid compared with a single copy plasmid [100].

Plasmid stability, one of the most important factor in recombinant processes, was examined next. Both pSYL101 and pSYL102 were unstable during repeated subculturing in antibiotic-free medium (Fig. 9, [20]). Two stable plasmid pSYL103 and pSYL104 (Fig. 10) were constructed by cloning the *parB* locus of

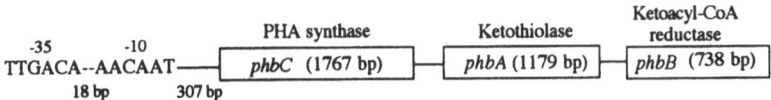

Fig. 8. *Alcaligenes eutrophus* PHA biosynthsis operon [6, 100]

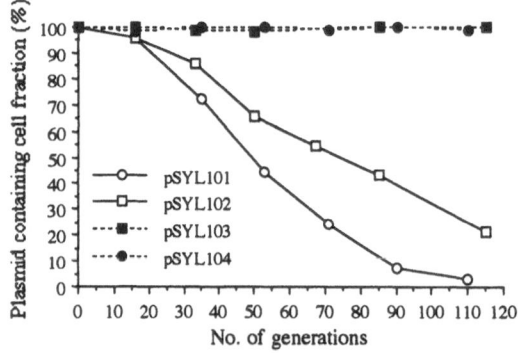

Fig. 9. Stability of plasmids pSYL101, pSYL102, pSYL103, and pSYL104 during the repeated subculturing in antibiotic-free medium [20]

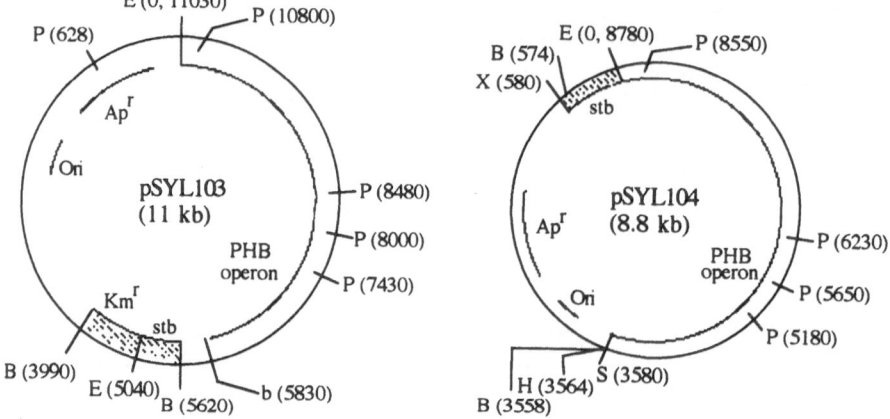

Fig. 10. Restriction maps of pSYL103 and pSYL104. Abbreviations are : *B, Bam*H1; *E,* EcoR I; *H, Hind* III; *P, Pst* I: *X, Xba* I, *b,* site of blunt end ligation; *stb, parB* locus of plasmid R1; *Km,* kanamycin; *Ap.* ampicillin; *r,* resistance gene; *Ori,* origin of replication. *Numbers in parenthesis* represent relative position of the restriction sites [20]

plasmid R1 [106] into pSYL102 and pSYL101, respectively. These two plasmids were 100% stable during repeated subculturing as shown in Fig. 9. This is important since the use of antibiotics is rarely possible in large scale industrial process. The importance of plasmid stability in PHA production has also been demonstrated by another group, who stabilized the plasmid using the *par* region of plasmid RP4 [107].

Host strain in recombinant processes should be carefully selected since growth rate, final achievable cell concentration, cell mass yield, substrate utilization, and byproduct formation vary considerably among different strains as demonstrated by Luli and Strohl [108]. We compared 15 *E. coli* strains, K12, B, W, and various derivatives, harboring another stable high copy number plasmid pSYL105 for its ability to synthesize and accumulate P(3HB). Cell growth, the rate of P(3HB) formation, P(3HB) content, P(3HB) yield, glucose utilization, and acetate

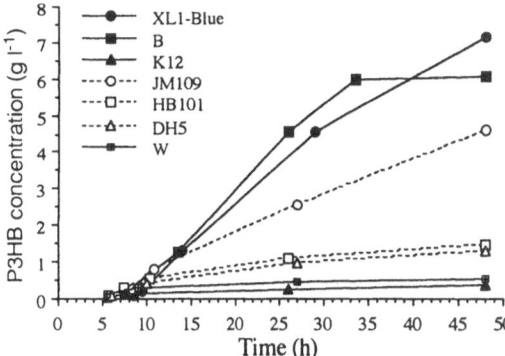

Fig. 11. Time profiles of P(3HB) concentration during flask cultures of seven recombinant *F. coli* strains harboring pSYL105 in LB containing 20 g l^{-1} glucose [127]

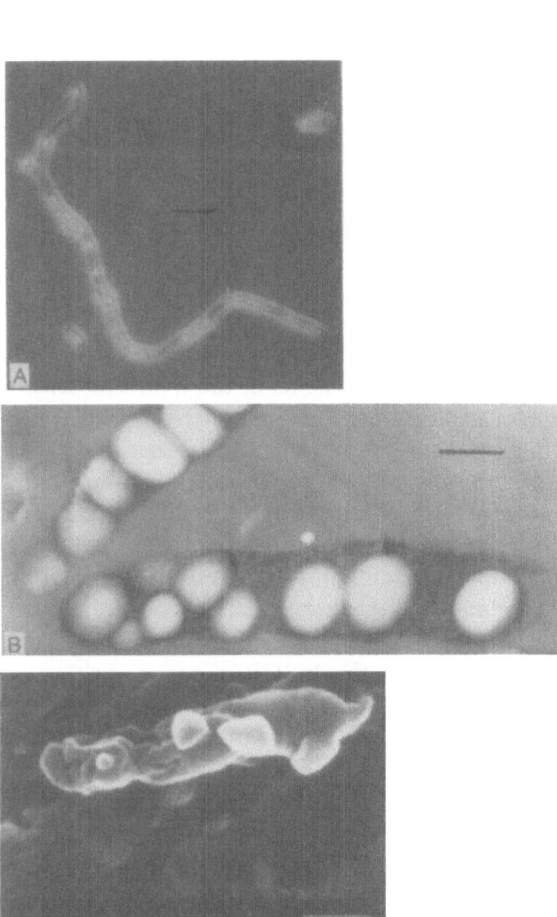

Fig. 12. Morphological changes of recombinant *E. coli* XL1-Blue (pSYL105) accumulating P(3HB); **A)** micrograph of an elongated cell accumulating P(3HB) granules (white bodies); **B)** TEM picture showing cells almost full of P(3HB); **C)** SEM picture of heavily distorted cell by protruding P(3HB) granules. *Bar* in (A) is 10 μm, while bars in (B) and (C) are 1 μm

formation varied significantly among the strains tested [127]. The time course of P(3HB) accumulation by seven *E. coli* strains harboring pSYL105 is shown in Fig. 11. XL1-Blue and B harboring pSYL105 synthesized P(3HB) at the fastest rate of 0.2 g P(3HB) per g true cell mass-hour, and accumulated P(3HB) up to $7 \, g \, l^{-1}$ after 48 h of flask culture in LB containing $20 \, g \, l^{-1}$ glucose. K12 (pSYL105) accumulated the least amount of P(3HB) with the lowest yield among the tested. Cells accumulating P(3HB) underwent considerable morphological changes, especially elongation (Fig. 12). Cells as long as 150 μm were observed under microscopy. The extent of elongation (filamentation) and the fraction of filamented cells also varied among the strains tested [127]. Cell elongation was found to be due to the inactivation of an essential cell division protein, FtsZ [109], either by the overexpression of the foreign PHA biosynthesis genes or by the accumulation of polymer granules. This was supported by the finding that overproduction of FtsZ in PHA accumulating cells of *E. coli* resulted in normally shaped (or slightly smaller) cells [128].

8.2 Production of Poly (3-hydroxybutyric acid) and Copolymer

E. coli has been the workhorse for the production of many recombinant DNA products, particularly proteins. There have been numerous reports on the high cell density culture of non-recombinant and recombinant *E. coli* strains (for reviews see [93,94.110]). One of the major problems in the high cell density culture of *E. coli* is the formation of acidic by-products, mainly acetic acid, which at high concentrations ($> 10 \, g \, l^{-1}$) inhibits growth and formation of re-combinant products [94,110]. Not to worry about acetate accumulation, we first carried out high cell density culture of XL1-Blue and HB101 harboring pSK2665 [99] using a membrane cell recycle reactor in complex medium containing glucose. In this system the medium was continuously taken out through an internal ceramic membrane module, and fresh medium was continuously supplemented (for the cell recycle system see [111]). We were able to obtain a cell concentration of more than $150 \, g \, l^{-1}$ with the P(3HB) content of 40–60% in 40 h ([101], unpublished results). Even though the membrane cell recycle system provides an easy way to obtain high density of *E. coli* cells by reducing the accumulation of acetic acid, it is not used practically in industry due to several problems including difficulty of scale-up and membrane fouling [42].

Fed batch culture has most often been employed for the production of many recombinant proteins with high productivity [41,94,110]. We examined a number of feeding strategies such as DO-stat, pH-stat, and specific growth rate control [41] for the production of P(3HB) by recombinant *E. coli*, and found the pH-stat method to be most suitable. The pH-stat fed batch culture of XL1-Blue (pSK2665) was carried out in an initial medium containing LB $+ 20 \, g \, l^{-1}$ glucose $+ 0.1 \, g \, l^{-1}$ ampicillin. Nutrient solution consisting of $400 \, g \, l^{-1}$ glucose $+ 100 \, g \, l^{-1}$ yeast extract $+ 100 \, g \, l^{-1}$ tryptone was added for a definite on-time

(50 ml as a pulse), whenever the culture pH became higher than 7.1 due to glucose depletion. Cell mass, P(3HB) concentration, P(3HB) content, and productivity of 117 g l^{-1}, 89 g l^{-1}, 76%, and 2.11 g l^{-1} h^{-1} P(3HB), respectively, were obtained in 42 h [102]. To avoid the use of expensive antibiotics, XL1-Blue harboring the stable high copy number plasmid pSYL104 was employed in the same culture system. Cell mass and P(3HB) concentrations of 101 and 81 g l^{-1}, respectively, were obtained in 39 h in antibiotic-free complex medium (Fig. 13, [20]). Cell mass increased linearly until 21 h. After 21 h the increase of cell mass was not due to the actual cell growth, but due to the accumulation of P(3HB), as shown by the constant (or slightly decreasing) true cell mass in Fig. 13. The P(3HB) content continuously increased to the final value of 80.1%. In both cultures mentioned above, acetate concentration never exceeded 6 g l^{-1}, well below the growth-inhibitory level, even though glucose was present at relatively high concentration compared with other recombinant *E. coli* processes [94]. This seems to be due to that the PHA biosynthesis enzymes efficiently utilized excess acetyl-CoA, which otherwise would form acetic acid.

Even though high concentration of P(3HB) was obtained as shown above, it is not economically feasible to use such a large amount of expensive yeast extract and tryptone. With the aim of lowering the medium cost, fed batch cultures with the same feeding strategy were carried out in a defined medium. Fed batch culture of XL1-Blue (pSYL104) in a defined medium resulted in the final cell mass, P(3HB) concentration, and P(3HB) content of 71.4 g l^{-1}, 16.3 g l^{-1}, and 22.8%, respectively, in 35 h [104]. Less accumulation of P(3HB) could be

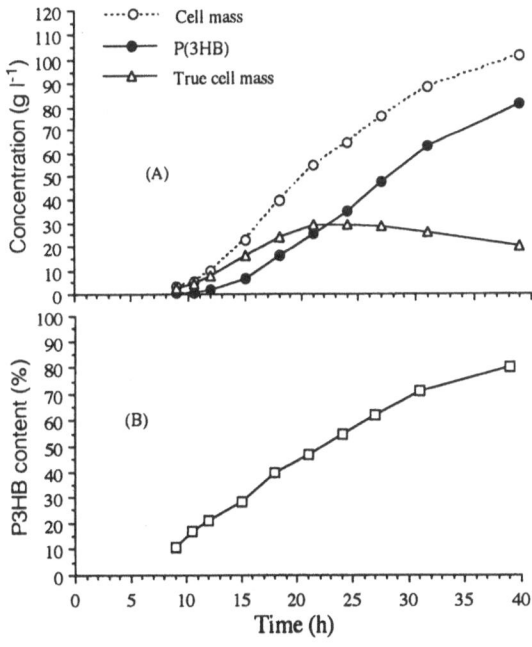

Fig. 13. The pH- stat fed batch culture of XL1-Blue (pSYL104) in complex medium: **A)** time profiles of cell mass, P(3HB), and true cell mass; **B)** time profile of P(3HB) content [20]

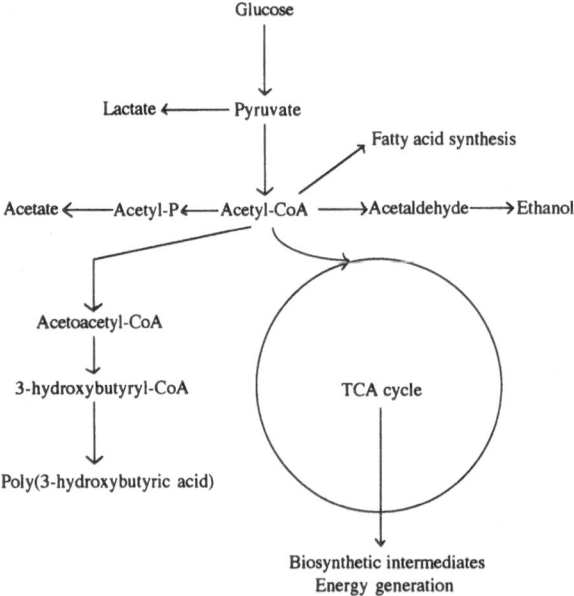

Fig. 14. Several competing metabolic pathways in recombinant *E. coli*

explained by the availability of acetyl-CoA as follows. The newly introduced PHA biosynthetic route has several competing metabolic pathways in *E coli* as shown in Fig. 14. Among these are formation of citric acid, acetic acid and ethanol, and fatty acid synthesis [112]. Therefore, the amount of acetyl-CoA available for P(3HB) synthesis is dependent on cellular metabolic status. In a complex medium major intermediary precursors including amino acids and vitamins are readily available for the synthesis of macromolecules, which results in the availability of abundant acetyl-CoA to the PHA synthetic pathway. The amount of acetyl-CoA available for P(3HB) synthesis is less in a defined medium since much more acetyl-CoA has to be used to synthesize biosynthetic precursors in other metabolic pathways. Regeneration of NADPH is also important since the second enzyme in PHA synthesis pathway, the reductase, requires it as a cofactor [5]. Synthesis of amino acids also required NADPH, and for example, as many as 8 moles of NADPH are required for the synthesis of 1 mole methionine from oxaloacetate [112]. This could be another reason for the lower accumulation of P(3HB) in a defined medium, in which all the amino acids had to be synthesized. It is also known that β-ketothiolase is inhibited by free CoA molecules [5]. Cells operate the TCA cycle intensively to meet the biosynthetic requirements in a defined medium, which releases free CoA molecules by the first enzyme in the cycle, citrate synthase. All of these seem to have resulted in a lower accumulation of P(3HB) in a defined or semi-defined medium compared with a complex medium. These were proven to be true by carrying out flask and fed batch cultures in a defined medium supplemented with one or several biosynthetic precursors such as amino acids and oleic acid [105, 129]. The concentration of P(3HB) increased by two to four fold when one or several amino acids were supplemented in a defined medium [129]. Amino acids that require more NADPH for synthesis generally resulted in higher P(3HB) accumulation. Oleic acid, which can be used as a precursor for fatty acid synthesis, was also found to promote P(3HB) synthesis. Furthermore, use of mutant strain of *E. coli* that is not able to produce acetic acid resulted in faster accumulation of P(3HB) especially when the medium contained more than 20 g l^{-1} glucose.

It was thought that the concentration of P(3HB) could be increased by supplementing small amount of complex nitrogen sources enough to support the cellular biosynthetic demands, and by developing a suitable cultivation strategy. We first examined 10 complex nitrogen sources at various concentrations for their ability to promote P(3HB) synthesis [105]. Some of the results are shown in Fig. 15. Supplementing 0.2% (wt/vol) tryptone, casamino acids, or casein hydrolysate promoted P(3HB) synthesis to a higher extent (up to 4 fold) than other complex nitrogen sources tested. Corn steep liquor, one or the cheapest complex nitrogen sources frequently used in industry, was also found to promote P(3HB) synthesis when supplemented together with yeast extract. By the addition of small amount of these complex nitrogen sources during the pH-stat fed batch cultures, either one or combination of two or three, more than 70 g l^{-1} of P(3HB) could be obtained in ca. 40 h. One of these results is shown as an

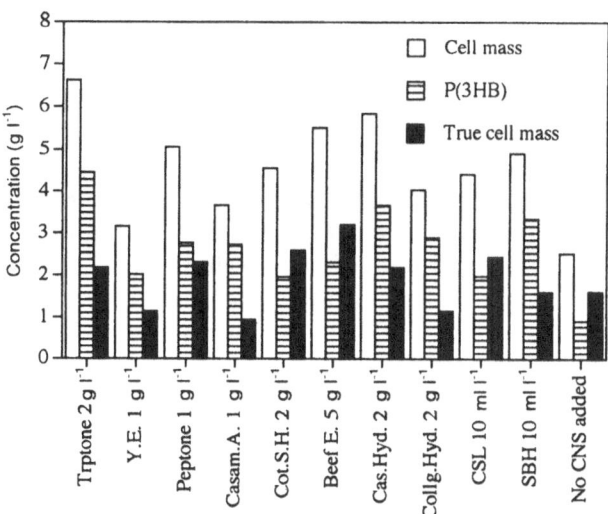

Fig 15. Effect of supplementing complex nitrogen sources on cell mass, P(3HB) concentration, and true cell mass. Abbreviations are: *Trptone*, Tryptone: *Y.E.*, yeast extract: *Casam. A.*, casamino acids; *Cot. S.H.*, cotton seed hydrolyzate; *Beef E.*, Beef extract; *Cas. Hyd.*, casein hydrolyzate; *Collg. Hyd.*, collagen hydrolyzate; *CSL*, corn steep liquor; *SBH*, soy bean hydrolyzate; No CNS, no complex nitrogen source supplemented [105]

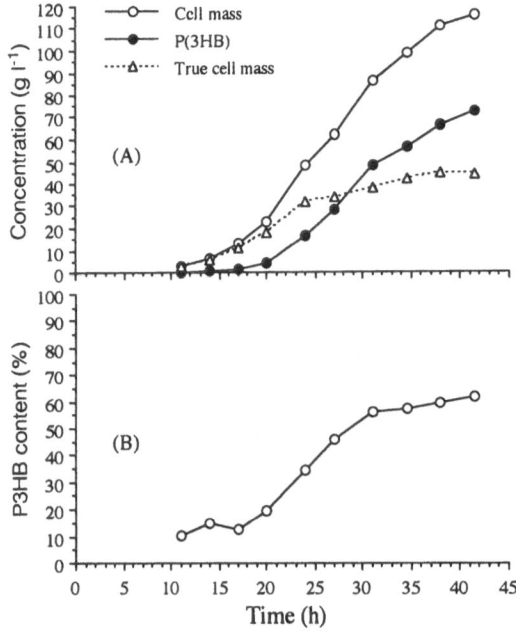

Fig. 16. Time profiles of **A** cell mass, P(3HB), true cell mass, and **B)** P(3HB) content during the fed batch culture of XL1-Blue (pSYL105) in R medium [103] supplemented with corn steep liquor and yeast extract [105]

example (Fig. 16). The initial medium contained per liter: 10 g glucose $+ 2$ g yeast extract $+ 5$ ml^{-1} corn steep liquor. The feeding solution contained per liter: 700 g glucose $+ 20$ g MgSO$_4$ 7H$_2$O $+ 8.8$ g yeast extract $+ 35$ m l^{-1} corn steep liquor. Upon the pH rise, 28.6 ml^{-1} of feeding solution was added as a pulse. Cell mass, P(3HB) concentration, and the P(3HB) content obtained in 41.5 h were 116 g l^{-1}, 72.2 g l^{-1}, and 62.2%, respectively (Fig. 16). This suggests that P(3HB) could be efficiently produced by supplementing a small amount of complex nitrogen sources in a defined medium.

E. coli W can utilize sucrose as a carbon source and this can lower the substrate cost considerably by using beet or cane molasses. We reported a strategy to carry out high cell density culture of *E. coli* W in a defined medium containing sucrose [103]. By the pH-stat fed-batch culture of *E. coli* W harboring pSYL104, cell mass and P(3HB) concentration of 124.6 and 34.3 g l^{-1}, respectively, were obtained in 48 h [103]. Again, the low concentration of P(3HB) could be increased by supplementing small amounts of complex nitrogen source (result not shown). Recently, Zhang et al. [113] also reported synthesis of P(3HB) by a newly isolated sucrose-utilizing strain of *E. coli* harboring the *A. eutrophus* PHA biosynthesis genes in flask cultures. Use of sucrose as a carbon source for the production of P(3HB) by recombinant *E. coli* seems to be promising, and waits further improvement.

Production of P(3HB) by recombinant *E. coli* has several advantages such as fast growth, use of various carbon substrates, large polymer accumulation, and the lack of depolymerases. Furthermore, purification of P(3HB) from recombinant *E. coli* seems to be much easier compared with other bacteria, since it was observed that cells accumulating large amount of P(3HB) became heavily distorted by protruding P(3HB) granules and wall fragility (Fig. 12). Expression of cloned lysis gene E of bacteriophage φX174 resulted in the release of P(3HB) granules into the medium [114]. However, the bacteriophage protein was not active in the stationary-phase, where most P(3HB) is accumulated [19]. Another system which involves the expression of T7 bacteriophage lysozyme gene was developed, and was shown to effectively release P(3HB) granules by adding a detergent at the end of culture [19]. A recent study by Y.K. Chang's group (KAIST, Taejon, Korea) showed that hypochlorite solution at low concentration could be used to recover P(3HB) from recombinant *E, coli* with negligible decrease of molecular weight probably due to the crystalline nature of P(3HB) in recombinant *E. coli* [130].

The copolymer P(3HB-*co*-3HV) was synthesized by recombinant *E. coli* LS5218 (*atoC fadR*) harboring the *A. eutrophus* PHA biosynthesis genes when propionic acid was also added [131]. The copolymer containing up to 40 mol % of 3-hydroxyvaleric acid could be obtained. Valeric acid was found to be better for the production of copolymer, even though it is more expensive than propionic acid (unpublished results). Since the pH-stat strategy was found to be not suitable for feeding nutrient solution containing glucose and propionic or valeric acid, we are currently developing a feeding strategy employing substrate concentration control for the production of P(3HB-*co*-3HV) to a high concentration.

Well-established physiology, genetics and molecular biology of *E. coli* allowed development of efficient strategies for the production and purification of PHA with several advantages described above. One major problem of employing recombinant *E. coli* for the production of PHA is use of pure oxygen to obtain high cell density. Even though lower cost of purification may counterbalance the cost of pure oxygen, much efforts are currently devoted to reduce the cellular oxygen demand without compromising the high PHA productivity. Also, cultivation strategy leading to high PHA accumulation in a defined medium is under development.

9 Other Microorganisms

There are numerous microorganisms, other than those described above, that are known to accumulate various PHAs (for review see [6]). Synthesis and accumulation of PHAs by 41 strains of phototrophic or chemilithotrophic bacteria have also been reported [115]. As mentioned already, most of these bacteria have not been employed in production studies due either to less accumulation of PHA or to the difficulty of high cell density cultivation. However, *Haloferax mediterrani*, an extremely halophilic aerobic chemorganotrophic archaeobacterium, has been considered as a PHA producer from starch. P(3HB) concentration, the P(3HB) content, and the yield of 6 g l^{-1}, 60%, and 0.33 g P(3HB) per g starch, respectively, were obtained from 2% starch [116]. It was later found that the polymer was actually the copolymer P(3HB-*co*-3HV) rather than P(3HB) homopolymer [117]. Due to the difficulty of being contaminated, continuous culture could be carried out for 3 months without strain degeneration. The steady-state PHA concentration was $1.5 \, g \, l^{-1}$ at a dilution rate of $0.02 \, h^{-1}$ [116], resulting in relatively low productivity. One more interesting microorganism to be mentioned is *Chromobacterium violaceum*, which was shown to accumulate a homopolymer of 3-hydroxyvaleric acid, P(3HV), up to 65% of dry cell weight from valeric acid as sole carbon source [118].

A number of recombinant bacteria other than *E. coli* have also been developed during cloning studies, and examined for the synthesis of various PHAs. Application of recombinant DNA techniques to the PHA synthesizing bacteria can be divided into two categories: introduction of substrate utilization genes and expression of PHA biosynthesis genes. Introduction of the β-galactosidase gene of Tn951 and the *E. coli gal* operon into *A. eutrophus* H16 allowed this strain to grow on lactose [119]. An attempt to make *A. eutrophus* H16 utilize sucrose by introducing the *Bacillus subtilis* levanase gene was not much successful either due to the poor secretion of the enzyme or poor sucrose uptake [120]. When the *A. eutrophus* PHA synthesis operon was introduced to *P. oleovorans*, the recombinant strain accumulated the blend of P(3HB) and P(3HH-*co*-3HO), which were stored in separate granules, from octanoic acid [121, 122]. It is interesting to see the formation or very large P(3HB) granules (diameter of 3.5 μm) in the recombinant *A. eutrophus* strain harboring the PHA synthase gene of *Rhodospirillum rubrum* [123]. The PHA synthases of *P. oleovorans* and *P. putita* were suggested to have different substrate specificity, since PHA formed were considerably different between the wild type *P. putita* and the PHA-negative mutant of *P. putita* harboring the *P. oleovorans* PHA synthase gene [124]. Among the many other recombinant strains constructed during cloning and characterization of PHA genes, it is most notable that a PHA-negative mutant of *P. putita* harboring the PHA synthase gene of *Thiocapsa pfennigii* was able to synthesize a polymer consisting of 3-hydroxybutyric acid (45.8 mol %), 3-hydroxyhexanoic acid (47.3 mol%), and 3-hydroxyoctanoic acid (4.2 mol %) from octanoic acid [125]. With the advances in cloning and molecular characterization of the PHA biosynthesis genes, recombinant microorganisms possessing ability to produce various PHAs (homopolymer, copolymer, or blend) with high productivity will be developed in the near future. It will be also beneficial to develop PHA producers that lack depolymerases by knocking out the genes.

10 Concluding Remarks

We have briefly reviewed production of PHA by a number of bacteria, and suggested advantages and disadvantages of employing them. Table 3 sum-

Table 3. Summary of PHA production by various microorganisms and culture methods

Bacterium	PHA	Culture method	Major substrates	Culture time h	Cell conc. g l⁻¹	PHA conc. g l⁻¹	PHA content %	Productivity g l⁻¹ h⁻¹	Reference
Alcaligenes eutrophus	P(3HB)	Glucose control fed batch	Glucose	49	124	92	74	1.87	[48]
Alcaligenes eutrophus	P(3HB)	Glucose control fed batch	Glucose	50	164	121	76	2.42	[48]
Alcaligenes eutrophus	P(3HB)	Recycled gas culture system	CO_2/H_2	40	85	61.5	72	1.54	[40]
Alcaligenes eutrophus	P(3HB)	Two-stage continuous	CO_2/H_2	–	–	20	–	0.9	[40]
Alcaligenes eutrophus	P(3HB)	Fed batch	Ethanol	50	63.5	47	74	0.94	[35]
Alcaligenes eutrophus	P(3HB/3HV)	Glucose control fed batch	Glucose + propionic acid	46	158	117	74	2.55	[49]
Alcaligenes eutrophus	P(3HB/3HV)	Glucose control fed batch	Glucose + propionic acid	39	113	64	56.5	1.64	[49]
Alcaligenes eutrophus	P(3HB/3HV)	Fed batch	Glucose + pentanoic acid	48	9.8	6.4	65	0.13	[3]
Alcaligenes latus	P(3HB)	Continuous	Sucrose	$D = 0.16\,h^{-1}$	–	16.2	–	2.6	T. Yamane
Alcaligenes latus	P(3HB/3HV)	One-stage continuous	Sucrose + propionic acid	$D = 0.15\,h^{-1}$	4.65	2	43	0.3	[126]
Azotobacter vinelandii	P(3HB)	Glucose control fed batch	Glucose + fish peptone	47	40.1	32	79.8	0.68	[64]
Azotobacter vinelandii	P(3HB/3HV)	Fed batch	Beet molasses + pentanoic acid	–	–	19–22	59–71	–	[63]
Haloferax mediterrani	P(3HB/3HV)	Continuous	Starch	$D = 0.02\,h^{-1}$	–	1.5	–	0.03	[116]
Klebsiella aerogenes	P(3HB)	Fed batch	Molasses	32	37	24	65	0.75	[113]
Paracoccus denitrificans	P(3HB/3HV)	Fed batch	Methanol + n-amyl alcohol	120	9	2.34	26	0.02	[16]

Table 3. (continued)

Bacterium	PHA	Culture method	Major substrates	Culture time h	Cell conc. g l^{-1}	PHA conc. g l^{-1}	PHA content %	Productivity g l^{-1} h^{-1}	Reference
Protomonas extorquens	P(3HB)	Fully automatic fed batch	Methanol	121	223	136	61	1.12	[15]
Protomonas extorquens	P(3HB)	Fully automatic fed batch	Methanol	170	233	149	64	0.88	[15]
Pseudomonas oleovorans	P(3HH/3HO)	Continuous	n-Octane	D = 0.09 h^{-1}	2.25	1.05	46.7	0.09	[90]
Pseudomonas oleovorans	P(3HH/3HO)	Continuous	n-Octane	D = 0.2 h^{-1}	11.6	2.9	25	0.58	[91]
Pseudomonas oleovorans	P(3HH/3HO)	Fed batch	n-Octane	38	37.1	12.1	33	0.32	[18]
Pseudomonas oleovorans	P(3HH/3HO)	Fed batch	Octanoic acid	45	41.8	15.5	37.1	0.34	unpublished results. [47]
Recombinant *E. coli*	P(3HB)	pH-stat fed batch	Glucose + LB medium	42	117	89	76	2.11	[102]
Recombinant *E. coli*	P(3HB)	pH-stat fed batch	Glucose	35	71.4	16.3	22.8	0.46	[104]
Recombinant *E. coli*	P(3HB)	pH-stat fed batch	Glucose + tryptone + thiamine	44	104.5	66.7	63.8	1.52	[105]
Recombinant *E. coli*	P(3HB)	pH-stat fed-batch	Glucose + yeast extract + corn steep liquor	41.5	116	72.2	62.2	1.74	[105]
Recombinant *E. coli*	P(3HB)	pH-stat fed-batch	Glucose + yeast extract + corn steep liquor + casein hydrolysate	41	112	71	72.3	1.98	unpublished results

marizes the production of PHAs with various microorganisms and culture methods. Much effort is currently being devoted to improve productivity, and there may be reports showing higher productivity by the time this paper is published. With all these efforts and the possibility of having an improved microorganism by recombinant DNA technology, economical production of PHA may become possible in the near future.

Acknowledgement. The authors are grateful to Dr. A. Steinbüchel, Dr. K. Gerdes, and Dr. B. Bachmann for kindly providing us with pSK2665, pKG1022, and LS5218, respectively. We thank Y.K. Lee, Dr. B. S. Kim, K. S. Yim, and Dr. Y. K. Chang for their participation in the various parts of the research and helpful discussion. Updated information provided by Dr. T. Yamane is greatly appreciated. Our work presented in this article was supported by the 'Korea Science and Engineering Foundation' and 'Korea Advanced Institute of Science and Technology'.

References

1. Byrom D (1987) TIBTECH 5 : 246
2. Holmes PA (1985) Phys. Technol. 16 : 32
3. Doi Y (1990) Microbial Polyesters. VCH Publishers, New York
4. Brandl H, Gross R, Lenz R, Fuller RC (1990) Adv. Biochem. Eng. Biotechnol. 41 : 77
5. Anderson AJ, Dawes EA (1990) Microbiol. Rev. 54 : 450
6. Steinbüchel A (1991) Polyhydroxyalkanoic acids. In: Byrom D (ed) Biomaterials Macmillan, London, p 123
7. King PP (1982) J. Chem. Technol. Biotechnol. 32 : 2
8. Utteley NL (1986) Proc. Biotech' 86, p. 171, Online Publications, Pinner
9. Inoue Y, Yoshie N (1992) Prog. Polym. Sci. 17 : 571
10. Steinbüchel A, Hustede E, Liebergesell M, Piper U, Timm A, Valentin H (1992) FEMS Microbiol. Rev. 103 : 217
11. Lafferty Rm, Korsatko B, Korsatko W (1988) Microbial Production of poly-β-hydroxybutyric acid. In Rehm HJ, Reed G (eds) Biotechnology, vol. 6b. VCH, Weinheim, p 136
12. Valentin HE, Lee EY, Choi CY, Steinbüchel A (1994) Appl. Microbiol. Biotechnol. 41 : 710
13. Hrabak O (1992) FEMS Microbiol. Rev. 103 : 251
14. Page WJ (1992) FEMS Microbiol. Rev. 103 : 149
15. Suzuki T, Yamane T, Shimizu S (1986) Appl. Microbiol. Biotechnol. 24 : 370
16. Ueda S, Matsumoto S, Takagi A, Yamane T (1992) Appl. Environ. Microbiol. 58 : 3574
17. Daniel M, Kim JH, Lebeault JM (1992) Appl. Microbiol. Biotechnol. 37 : 702
18. Preusting H, Houten R, Hoefs A, Langenberghe E, Favre-Bulle O, Witholt B (1993) Biotechnol. Bioeng. 41 : 550
19. Fidler S, Dennis D (1992) FEMS Microbiol. Rev. 103 : 231
20. Lee SY, YIM KS, Chang HN, Chang YK (1994) J. Biotechnol. 32 : 203
21. Doi Y, Segawa A, Kawaguchi Y, Kunioka M (1990) FEMS Microbiol. Lett. 67 : 165
22. Bradel R, Kleinke A, Reichert K (1989) Proc. DECHEMA Biotechnology Conferences 3, VCH, Weinheim, p 207
23. Yamane T (1993) Biotechnol. Bioeng. 41 : 165
24. Yamane T (1992) FEMS Microbiol. Rev. 103 : 257
25. Dawes EA (1990) Novel Microbial Polymers. In: Dawes EA (ed) Novel biodegradable microbial polymers. Kluwer, Dordrecht, p 3

26. Holmes PA, Wright, LF, Collins SH (1981) European Patent 0052459
27. Doi Y, Segawa A, Nakamura S, Kunioka M (1990) Production of biodegradable copoolyesters by *Alacaligenes eutrophus*. In: Novel Biodegradable Microbial Polymers (ed. Dawes EA), p. 37, Kluwer, Dordrecht
28. Baptist JN (1959) US Patent 3036959
29. Baptist Jn (1960) US Patent 3044942
30. Byrom D (1991) Miscellaneous Biomaterials. In: Biomaterials (ed. Byrom D), p. 333, Macmillan, London
31. Byrom D (1992) FEMS Microbiol. Rev. 103:247
32. Steinbüchel A (1992) Curr. Opin. Biotechnol. 3:291
33. Laferty RM (1979) US Patent 4138291
34. Senior PJ, Collins SH, Richardson KR (1986) European Patent 204442
35. Alderete JE, Karl DW, Park CH (1993) Biotechnol. Prog. 9:520
36. Schlegel HG, Gottschalk G, Bartha R (1961) Nature 191:463
37. Repaske R, Mayer R (1976) Appl. Environ. Microbiol. 32:592
38. Ishizaki A, Tanaka K (1990) J. Ferment. Bioeng. 69:170
39. Ishizaki A, Tanaka K (1991) J. Ferment, Bioeng. 71:254
40. Ishizaki A, Tanaka K (1992) Abstracts of International Symposium on Bacterial Polyhydroxyalkanoates' 92, P117, Göttingen
41. Yamane T, Shimizu S (1984) Adv. Biochem. Eng. Biotechnol. 30:147
42. Chang HN, Furusaki S (1991) Adv. Biochem. Eng. Biotechnol. 44:27
43. Lee YW, Yoo YJ (1991) Korean J. Appl. Microbiol. Biotechnol. 19:186
44. Suzuki T, Yamane T, Shimizu S (1986) J. Ferment. Technol. 64:317
45. Luli GW, Schlasner SM, Ordaz DE, Mason M, Strohl WR (1987) Biotechnol. Techniq. 1:225
46. Park YS, Kai KK, Ijima S, Kobayashi T (1992) Biotechnol. Bioeng. 40:686
47. Chang HN, Lee SY (1994) Abstracts of Advances in Biopolymer Engineering, p. 15, Palm Coast, FL
48. Kim BS, Lee SC, Lee SY, Change HN, Chang YK, Woo SI (1994) Biotechnol. Bioeng. 43:892
49. Kim BS, Lee SC, Lee SY, Chang HN, Chang YK, Woo SI (1994) Enz. Microbiol. Technol. 16:556
50. Hangi UJ (1990) Pilot scale production of PHA with Alcaligenes latus. In: Novel Biodegradable Microbial Polymers (ed. Dawes EA), p. 65, Kluwer Publishers, Dordrecht
51. Lafferty RM, Braunegg G (1984) European Patent 144017
52. Lafferty RM, Braunegg G (1984) European Patent 149744
53. Lafferty RM, Braunegg G (1988) US Patent 4786598
54. Lafferty RM, Braunegg G (1990) US Patent 4957861
55. Stockadale H, Ribbons DW, Dawes EA (1968) J. Bacteriol. 95:1798
56. Ritchie GAF, Dawes EA (1969) Biochem. J. 112:803
57. Ritchie GAF, Senior PJ, Dawes EA (1971) Biochem. J. 121:308
58. Senior PJ, Dawes EA (1973) Biochem. J. 134:225
59. Page WJ, Knosp O (1989) Appl. Environ. Microbiol. 55:1334
60. Page WJ, Knosp O (1992) US Patent 5096819
61. Page WJ (1989) Appl. Microbiol. Biotechnol. 31:329
62. Page WJ (1992) Appl. Microbiol. Biotechnol. 38:117
63. Page WJ, Manchak J, Rudy B (1992) Appl. Environ. Microbiol. 58:2866
64. Page WJ, Cornish A (1993) Appl. Environ. Microbiol. 59:4236
65. Powell KA, Collins BA (1982) US Patent 4336334
66. Suzuki T, Yamane T, Shimizu S (1986) Appl. Microbiol. Biotechnol. 23:322
67. Suzuki T, Yamane T, Shimizu S (1986) Appl. Microbiol. Biotechnol. 24:366
68. Suzuki T, Deguchi H, Yamane T, Shimizu S, Gekko K (1988) Appl. Microbiol. Biotechnol. 27:487
69. Ueda S, Matsumoto S, Takagi A, Yamane T (1992) FEMS Microbiol. Lett. 98:57
70. Kang CK, Lee HS, Kim JH (1993) Biotechnol. Lett. 15:1017
71. Bourque D, Ouellette B, Andre G, Groleau D (1992) Appl. Microbiol. Biotechnol. 37:7
72. Huisman GW, Deleeuw O, Eggink G, Witholt B (1989) Appl. Environ. Microbiol. 55:1949
73. Haywood GW, Anderson AJ, Dawes EA (1989) Biotechnol. Lett. 11:471
74. DeSmet MJ, Eggink G, Witholt B, Kingma J, Wynberg H (1983) J. Bacteriol. 154:870
75. Brandl H, Gross RA, Lenz RW, Fuller RC (1988) Appl. Enviorn. Microbiol. 54:1977
76. Lageveen RG, Huisman GW, Preusting H, Ketelaar P, Eggink G, Witholt B (1988) Appl. Environ. Microbiol. 54:2924

77. Timm A, Steinbüchel A (1990) Appl. Environ. Microbiol. 56:3360
78. Haywood GW, Anderson AJ, Ewing EF, Dawes EA (1990) Appl. Environ. Microbiol. 56:3354
79. Huijberts GNM, Eggink G, DeWaard P, Huinsman GW, Witholt B (1992) Appl. Environ. Microbiol. 58:536
80. Huijiberts GNM, DeRijk TC, DeWaard P, Eggink G (1994) J. Bacteriol. 176:1661
81. Steinbüchel A, Wiese S (1992) Appl. Microbiol. Biotechnol. 37:691
82. Timm A, Wiese S, Steinbüchel A (1994) App. Microbiol. Biotechnol. 40:669
83. Ramsay BA, Saracovan I, Ramsay JA, Marchessault RH (1992) Appl. Environ. Microbiol. 58:744
84. Saito Y, Doi Y (1993) Int. J Biol. Macromol. 15:287
85. Lenz RW, Kim YB, Fuller RC (1992) Biotechnol. Lett. 14:445
86. Kimura H, Yoshida Y, Doi Y (1992) Biotechnol. Lett. 14:445
87. Bertrandt JL, Ramsay BA, Ramsay JA, Chavarie C (1990) Appl. Environ. Microbiol. 56:3133
88. Gagnon KD, Lenz RW, Farris RJ (1992) Rubber Chem. Technol. 65:761
89. Witholt B, Lageveen RG (1992) US Patent 5135859
90. Preusting H, Kingma J, Witholt B (1991) Enzyme Microb. Technol. 13:770
91. Presuting H, Hazenberg W, Witholt B (1993) Enzyme Microb. Technol. 15:311
92. Ramsay BA, Saracovan I, Ramsay JA, Marchessault RH (1991) Appl. Environ. Microbiol. 57:625
93. Shatzman AR (1990) Curr. Opin. Biotechnol. 1:5
94. Yee L, Blanch HW (1992) Bio/Technol. 10:1550
95. Schubert P, Steinbüchel A, Schlegel HG (1988) J. Bacteriol. 170:5837
96. Slater SC, Voige WH, Dennis DE (1988) J. Bacteriol. 170:4431
97. Peoples OP, Sinskey AJ (1989) J Biol. Chem. 264:15293
98. Peoples OP, Sinskey AJ (1989) J. Biol. Chem. 264:15298
99. Schubert P, Kruger N, Steinbüchel A (1991) J. Bacteriol. 173:168
100. Janes B, Hollar J, Dennis De (1990) Molecular characterization of the poly-β-hydroxybutyrate biosynthetic pathway of *Alacaligenes eutrophu* H16. In: Novel Biodegradable Microbial Polymers (ed. Dawes EA), p. 175, Kluwer Academic Publishers, Dordrecht
101. Kim BS, Lee SY, Chang HN (1992) Abstracts of International Symposium on Bacterial Polyhydroxyalkanoates '92, Göttingen, p. 120
102. Kim BS, Lee SY, Chang HN (1992) Biotechnol. Lett. 14:811
103. Lee SY, Chang HN (1993) Biotechnol. Lett. 15:971
104. Lee SY, Chang HN, Chang YK (1994) Ann. NY Acad. Sci. 721:43
105. Lee SY, Chang HN (1994) J. Environ. Polymer Degrad. 2:169
106. Gerdes K (1988) Bio/Technol. 6:1402
107. Haigermoser C, Chen GQ, Gorhmann E, Hrabak O, Schwab H (1993) J. Biotechnol. 28:291
108. Luli GW, Strohl WR (1990) Appl. Environ. Microbiol. 56:1004
109. Lutkenhaus J (1990) Trends Genet 6:22
110. Riesenberg D (1991) Curr. Opin. Biotechnol. 2:380
111. Kang BC, Lee SY, Chang HN (1993) Biotechnol. Bioeng. 42:1107
112. Neidhardt FC, Ingraham JL, Schaechter M (1990) Physiology of the bacterial cell: A molecular approach. Sinauer Associates Sunderland, MA
113. Zhang H, Obias V, Gonyer K, Dennis D (1994) Appl. Environ . Microbiol. 60:1198
114. Busse HJ, Kalousek S, Lubitz W (1992) Abstracts of Internatinal Symposium on Bacterial Polyhydroxyalkanoates '92, Göttinger, p 273
115. Liebergesell M, Hustede E, Timm A, Steinbüchel A, Fuller RC, Lenz RW Schlegel HG (1991) Arch. Microbiol. 155:415
116. Garcia Lillo JA, Rodriguez-Valera F (1990) Appl. Environ. Microbiol. 56:2517
117. Rodriguez-Valera F, Garcia Lillo JA (1990) Halobacteria as producers of poly-β-hydroxy-alkanoats. In: Novel Biodegradable Microbial Polymers (ed. Dawes EA), p. 425, Kulwer Academic Publishers, Dordrecht.
118. Steinbüchel A, Debzi EM, Marchessault RH, Timm A (1993) Appl. Microbiol. Biotechnol. 39:443
119. Pries A, Steinbüchel A, Schlegel HG (1990) Appl. Microbiol. Biotechnol. 33:410
120. Fries K, Lafferty RM (1989) J. Biotechnol. 10:285
121. Timm A, Byrom D, Steinbüchel A (1990) Appl. Microbiol. Biotechnol. 33:296
122. Preusting H, Kingma, J, Huismann G, Steinbüchel A, Witholt B (1993) J. Environ. Polym. Degrad. 1:11
123. Hustede E, Steinbüchel A, Schlegel HG (1992) FEMS Microbiol. Lett. 93:285

124. Huisman GW, Wonink E, De Koning G, Preusting H, Witholt B (1992) Appl. Microbiol. Biotechnol. 38:1
125. Liebergesell M, Mayer F, Steinbüchel A (1993) Appl. Microbiol. Biotechnol. 40:292
126. Ramsay B, Lomaliza K, Chavaric C, Dube B, Bataille P, Ramsay J (1990) Appl. Envron. Microbiol. 56:2093
127. Lee SY, Lee KM, Chang HN, Steinbüchel A (1994) Biotechnol. Bioeng. 44:1337
128. Lee SY (1994) Biotechnol. Lett. 16:1247
129. Lee SY, Lee YK, Chang HN (1995) J. Ferment. Bioeng. in press
130. Hahn SK, Chang YK, Lee SY (1995) Appl. Environ. Microbiol. in press
131. Slater S, Gallaher T, Dennis D (1992) Appl. Environ. Microbiol. 58:1089

The Potential of Using Cyanobacteria in Photobioreactors for Hydrogen Production

S. A. Markov, M. J. Bazin and D. O. Hall
Environmental Biotechnology Research Group, Division of Life Sciences, King's College London, Camden Hill Road, London, W8 7AH, UK

This review surveys data on cyanobacterial hydrogen photoproduction with a view to a use of cyanobacteria in photobioreactors for hydrogen production.

Three groups of cyanobacteria can be distinguished based on their hydrogen producing physiological characteristics: a) heterocystous filamentous, b) nonheterocystous filamentous and c) nonheterocystous unicellular strains. Information on certain metabolically unique strains of cyanobacteria which have special promise for practical hydrogen production is reviewed.

Advances in Biochemical Engineering
Biotechnology, Vol. 52
Managing Editor: A. Fiechter
© Springer-Verlag Berlin Heidelberg 1995

Environmental parameters and physiological factors which might be of use to optimize cyanobacterial hydrogen generation are identified and summarized. These parameters include: light intensity, gas atmosphere (CO_2, N_2 and O_2), temperature, pH, carbohydrate substrates, metal ions, H_2 uptake systems, age of cyanobacterial culture, cell density, and immobilization of cells.

Immobilization of cyanobacterial cells in polyurethane or polyvinyl foams, hollow fibres, glass beads and cotton, leads to an increase/or stabilization for several months or more of hydrogen production. Cell immobilization results in an increased heterocyst frequency (up to 40%). Immobilized cells in comparison with free living cells are likely to be more suitable for hydrogen production in photobioreactors.

Nitrogenase is a major catalytic enzyme for hydrogen production in cyanobacteria, which can express three distinct nitrogenases: molybdenum nitrogenase, vanadium nitrogenase and iron nitrogenase. Details of the biochemical manipulation of nitrogenase-catalyzed hydrogen photoproduction is reviewed. In the case of molybdenum-independent nitrogenases, hydrogen production is significantly higher than for molybdenum-containing nitrogenase. Cyanobacterial hydrogenase is an enzyme which catalyzes both hydrogen evolution and hydrogen uptake. Several means to inhibit hydrogenase uptake activity are described. Removal of H_2 from the cyanobacterial cell environment in a continuous-flow photobioreactor is an effective method of preventing hydrogen uptake. Cyanobacteria may also be genetically modified for deletion of the uptake hydrogenase gene. Other objectives of genetic research related to cyanobacterial hydrogen photoproduction are reviewed.

A number of reports on cyanobacterial photobioreactors for hydrogen production are described. Such photobioreactors can be operated continuously for several months or more. Under outdoor conditions H_2-producing photobioreactors operated at good efficiencies. For example a 5 l outdoor photobioreactor with *Oscillatoria* sp. produced hydrogen for a week at rates of $8.2\ \mathrm{ml\,g^{-1}\,h^{-1}}$.

Particular attention in the review is given to fixed bed hollow fibre column photobioreactors. Several advantages of those constructions are characterized. Photoproduction of hydrogen at rates up to $200\ \mathrm{ml\,g^{-1}\,h^{-1}}$ in a photobioreactor with hollow fibre immobilized *Anabaena variabilis* was observed for more than 5 months. An example of a computer-controlled equipment system for cyanobacterial growth is given.

1 Introduction

Hydrogen is considered to be an environmentally desirable fuel since its combustion product (water) is non-polluting and it can be produced in renewable energy systems. There are currently several industrial methods for production of H_2 mostly from natural gas, oil, coal and water. Nearly 90% of H_2 is obtained by the reaction of natural gas or light oil with steam at high temperature (reforming). Coal gasification and electrolysis of water are other industrial methods for H_2 production. These industrial methods mainly consume fossil energy sources and sometimes hydroelectricity.

Considerable research has been done on the utilization of solar energy for H_2 production. A number of reviews have been published in the past on H_2 production by water splitting using photoelectrochemical, photochemical and photobiological methods [1–4]. Photochemical methods use a photosensitizer to promote water photolysis, but the yields are poor. Photoelectrochemical methods apply semi-conductor electrodes for light absorption and charge separation. The problem here is to find suitable electrode materials with a high enough band gap to generate the photovoltage required for water splitting. Photobiological H_2 production has a number of advantages and capitalizes on the fact that many microbial species produce molecular hydrogen. It has been suggested by several investigators that the most suitable candidates for the development of an environmentally acceptable technology for hydrogen production are cyanobacteria [5–9]. This is because cyanobacteria are unique in their ability to produce hydrogen using water as their ultimate electron substrate and solar energy as an energy source. Simultaneously with H_2 evolution cyanobacteria consume CO_2 and thus may also provide a mechanism for CO_2 removal from air.

The design, optimization and practical demonstration of computer-controlled photobioreactors in which solar energy is used for hydrogen production by cyanobacterial cells would be an important step toward an advanced hydrogen production technology. Development of photobioreactors is a rapidly developing branch of environmental biotechnology based on the utilization of light energy and wasted CO_2 [10–14].

In this review we concentrate on the physiological, biochemical and genetic characteristics of cyanobacteria in relation to the application of cyanobacteria in photobioreactors for hydrogen production.

2 Physiology of Hydrogen Production by Cyanobacteria

2.1 The Cyanobacteria

Cyanobacteria (also called blue-green algae) are an ancient group of organisms, which in evolution connect heterotrophic bacteria to the higher plants. Like all bacteria, they lack nuclei, mitochondria and chloroplasts. However, their O_2-evolving and CO_2-consuming photosynthesis comprises two photosystems that generate reductants from water in mechanisms similar to those of green plants. Having features common to both bacteria and plants, cyanobacteria are highly flexible in their metabolism.

Many cyanobacterial species can also fix molecular nitrogen from air and convert it to organic nitrogen-containing compounds such as proteins, pigments, etc. This is a major reason why they are able to grow autotrophically using carbon dioxide, nitrogen from air, and water, with a simple inorganic nutrient supply and only using light as an energy source. Nitrogen-fixing cyanobacteria have the simplest nutritional requirements of all organisms.

Although cyanobacteria are typically photosynthetic, some of the species can grow in the dark in the presence of organic substrates [15–18]. Some cyanobacteria can also use sulphide as an electron donor for photosynthesis and live under anaerobic conditions [19].

Thus the cyanobacteria are highly adaptive to different environmental conditions. They live practically everywhere: in salt and fresh water, deserts, hot springs (at temperatures up to 73 °C) and in the Antarctic. Cyanobacteria are also found as symbionts in other organisms such as lichens, cycads and water ferns.

Different aspects of cyanobacterial physiology, biochemistry and ecology have been reviewed in number of books [20–28].

Cyanobacteria have also been considered by several workers to be a potentially excellent source of chemicals and fuels and also feed for animals and human consumption [5, 29, 30].

Under suitable conditions cyanobacteria evolve molecular hydrogen. The first observation of this phenomenon was reported in 1896, but the cultures used were probably not pure [31]. Confirmatory evidence that cyanobacteria do produce hydrogen came much later [32, 33]. The work was performed with a heterocystous *Anabaena* and hydrogen production was observed under an atmosphere of argon in the absence of molecular nitrogen.

Most studies of hydrogen production have been carried out with nitrogen-fixing cyanobacteria. It was shown that nitrogenase (the enzyme responsible for nitrogen fixation) is a major catalyst of hydrogen production in cyanobacteria [34] according to (1):

$$8H^+ + N_2 + 8e^- + 12ATP \rightarrow 2NH_3 + H_2 + 12ADP + 12P_i. \tag{1}$$

A so-called reversible hydrogenase catalyzes hydrogen production in cyanobacteria, although the rate of this production is low compared to the nitrogenase-catalyzed reaction [35].

2.2 Nitrogenase-Catalyzed Hydrogen Evolution

Cyanobacteria with nitrogenase-catalysed hydrogen production can be classified into three groups based on their morphological and physiological characteristics. These are: heterocystous, nonheterocystous filamentous and nonheterocystous unicellular species.

2.2.1 Heterocystous Cyanobacteria

Nitrogenase and reversible hydrogenase are very sensitive to molecular oxygen [36]. Among hydrogen-evolving organisms only cyanobacteria have the unique ability to carry out both oxygen-evolving photosynthesis and hydrogen evolution within the same organism.

In heterocystous strains specialised cells, heterocysts, are the site of nitrogenase reactions under aerobic growth conditions (Fig. 1). Comprehensive information on different aspects of heterocysts is available from several monographs and review articles [22, 37–40]. Many heterocystous strains have been studied for hydrogen production. Among these are: *Anabaena cylindrica*, *A. azollae*, *A. variabilis*, *A. flos-aquae*, *Chlorogloeopsis fritschii*, *Mastigocladus laminosus*, *M. thermophilus*, *Nostoc muscorum*, *Nostoc* sp. [41]. Heterocysts possess a number of morphological and biochemical modifications designed to protect nitrogenase from oxygen inactivation. Heterocysts have a modified photosynthetic mechanism lacking both photosynthetic carbon dioxide fixation and oxygen evolution. On the other hand, heterocysts possess all the necessary Photosystem I components and are capable of photophosphorylation with synthesis of ATP. The Photosystem I activity provides at least part of the required energy for nitrogenase-catalyzed hydrogen evolution.

Heterocysts show high rates of endogenous respiration [42]. According to Kangatharalingam et al. [43] heterocysts can form aggregates and create an environment of low partial pressure of O_2 where complete photosynthesis is not operative.

The source of electrons (reductant) for nitrogenase activity in heterocystous cyanobacteria is fixed carbon–carbohydrates or even amino acids [44]. Fixed carbon imported from vegetative cells is probably via the pore channels which connect heterocysts and vegetative cells. Using radioactive carbon it was shown that maltose, which is synthesised in vegetative cells from carbon dioxide and water, was the first radioactive compound to appear in the heterocysts [45]. Addition of different carbohydrates (erythrose, fructose and glucose) to isolated heterocysts of *A. cylindrica* and *A. variabilis* stimulated nitrogenase activity [46].

VEGETATIVE CELL   HETEROCYST

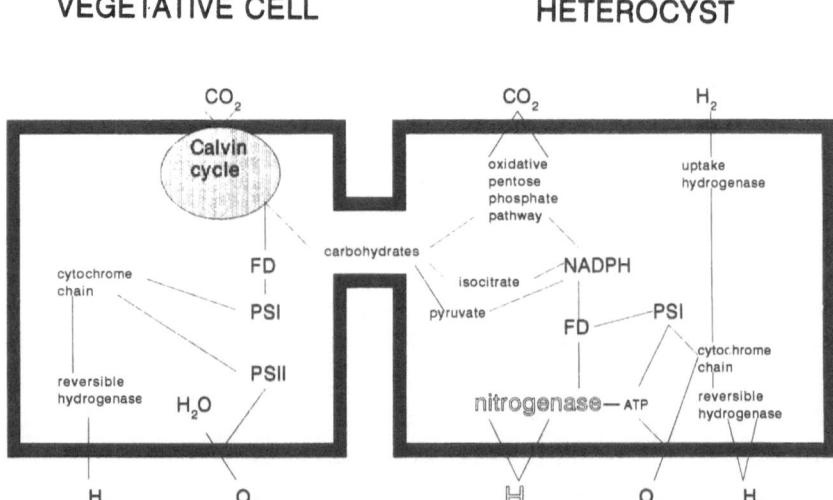

Fig. 1. Diagrammatic relationship between hydrogen production and cell metabolism in heterocystous cyanobacteria

Utilization of carbohydrates with release of CO_2 occurs primarily through the oxidative pentosephosphate pathway [17]. The activity of the two enzymes of this pathway, glucose 6-phosphate dehydrogenase and 6-phosphogluconate dehydrogenase, are 7–8-times higher in heterocysts than in vegetative cells [47]. The NADPH produced by the oxidative pentosephosphate pathway donates electrons to nitrogenase indirectly via ferredoxin: NADP oxidoreductase and ferredoxin [42]. The immediate electron source to nitrogenase is probably ferredoxin [48]. During iron starvation flavodoxin can replace ferredoxin [49–51].

Some reductants can serve as sources of electrons to ferredoxin in heterocysts. These include:

– NADH generated by the glycolytic pathway [52].

– isocitrate by means of isocitrate dehydrogenase. High rates of isocitrate dehydrogenase activity have been observed in heterocysts [42, 53]. This reaction can generate reduced ferredoxin.

– Pyruvate may provide reduced ferredoxin via the enzyme pyruvate: ferredoxin oxidoreductase [42].

– Molecular hydrogen donates electrons by means of uptake hydrogenase to ferredoxin via the photosynthetic electron transfer chain in light or the respiratory chain (so-called oxyhydrogen reaction) in heterocyst [53].

The source of reductant depends on the environmental conditions.

The immediate source of energy for nitrogenase-catalyzed hydrogen production is ATP. ATP is generated by photosynthesis or respiration.

The number of heterocysts in cyanobacterial cultures may vary depending on a number of factors. Normally, about 5 to 10% of the cells in the filaments develop into heterocysts, following removal of fixed (combined) nitrogen from the medium [54]. 7-azotryptophan [55, 56] and carbon compounds [57] stimulate the differentiation of heterocysts in cyanobacteria. Frequencies of heterocysts may reach as high as 30–60% in symbiotic *A. azollae* growing in leaf cavities of the water fern *Azolla* [58–60]. Immobilization of *A. azollae* cells in polyvinyl and polyurethane foams leads to an increase in heterocyst frequency of up to 18% [61].

Nitrogenase activity has also been detected in vegetative cells under anaerobic or microaerobic conditions [39, 62]. However, to date, hydrogen production by vegetative cells has not been observed.

2.2.2 Nonheterocystous Unicellular Cyanobacteria

Hydrogen production in nonheterocystous unicellular strains is always under the influence of O_2, produced during photosynthesis. There is no universal system for the oxygen protection of nitrogenase-catalysed hydrogen production in nonheterocystous cyanobacteria. Temporal separation of photosynthesis and hydrogen production appears to occur in most cases. A temporal separation of nitrogenase activity and photosynthesis in nonheterocystous cyanobacteria has been studied in detail by Gallon and co-workers [63, 64]. When *Gloeothece* cells were incubated under a 12-h light-12-h dark regime, under conditions that resemble the natural diurnal cycle, significant nitrogenase activity was detected only during the dark period. Nitrogenase activity was apparently supported by the catabolic breakdown of reserve carbohydrate accumulated in the previous light period [63, 65, 66].

Most unicellular cyanobacteria can produce hydrogen under anaerobic or microaerobic conditions [67]. On the other hand, only a few such cyanobacteria were shown to be capable of aerobic hydrogen production. These include *Gloeothece* [68, 69] and *Synechoccoccus* sp. Miami BG 043511 [70]. The unicellular cyanobacterium, *Synechococcus* can produce hydrogen in the presence of oxygen but does not exhibit hydrogen uptake activity [70]. This strain produced oxygen simultaneously with hydrogen and does not release carbon dioxide. This indicates that electrons from water are efficiently donated to nitrogenase for hydrogen production. The mechanism for the simultaneous photoproduction of hydrogen and oxygen by this cyanobacterium is not fully understood.

2.2.3 Nonheterocystous Filamentous Cyanobacteria

Among nonheterocystous filamentous strains, hydrogen photoproduction has been most intensively studied in the marine cyanobacteria *Lyngbya* sp.

Isolate N. 108 [71, 72], *Oscillatoria* sp. Miami BG 7 [73, 74], and *Phormidium valderianum* [75, 76]. Greater amounts of hydrogen photoproduction was shown in *Oscillatoria* sp. compared to the heterocystous cyanobacterium *A. cylindrica* (2–2.5 times more) [73]. *Oscillatoria* sp. Miami was shown to exhibit sustained and high rates of hydrogen photoproduction via a two-step process of aerobic photosynthesis and anaerobic hydrogen photoproduction. The maximum H_2 rate observed so far ($12 \, \mathrm{ml \, g^{-1} \, h^{-1}}$) was maintained for one week. In the first step, cells were cultured in a fixed-nitrogen-limited medium. During this period, the concentration of glycogen (the hydrogen donating substrate) increased. This created a high C/N ratio in the cells and resulted in enhancement of nitrogenase synthesis. On the other hand, the concentration of photosynthetic pigments such as chlorophyll and phycobilin decreased thus minimizing oxygen inhibition of nitrogenase (or hydrogenase). During the second step, the cells were incubated under an argon atmosphere. Upon illumination, glycogen was hydrolyzed to glucose which was converted to hydrogen and carbon dioxide [77].

The reactions are:
Step 1: Aerobic

$$12H_2O + 6CO_2 = C_6H_{12}O_6 + 6H_2O + 6O_2 \tag{2}$$

$$xC_6H_{12}O_6 = \text{glycogen} + yH_2O$$

Step 2: Anaerobic

$$\text{Glycogen} + yH_2O = xC_6H_{12}O_6 \tag{3}$$

$$C_6H_{12}O_6 + 6H_2O = 6CO_2 + 12H_2$$

Hydrogen photoproduction by *Lyngbya* was observed under anaerobic conditions for 7 d with simultaneous consumption of cell reserve carbohydrates [71]. Under hydrogen photoproduction conditions (N-poor medium without fixed nitrogen, anaerobic conditions) the degradation of Photosystem II proteins of *Lyngbya* sp. was observed [72]. The transfer of *Lyngbya* from growing conditions to hydrogen-producing conditions led to the loss of oxygen evolution within a day, followed by the loss of the reaction centre in PSII within 3 d. Phycocyanin decreased significantly and the capability of energy transfer from PSII to PSI was lost.

2.3 Hydrogenase-Catalyzed Hydrogen Evolution

Reactions involving reversible hydrogenase in cyanobacteria still need much more investigation. For heterocystous cyanobacteria, hydrogenase activity is usually detectable in ammonia or nitrate-grown cells where nitrogenase is repressed and it is possible to distinguish nitrogenase and hydrogenase activities [78]. The physiological role of reversible hydrogenase remained unclear until recently. There is now evidence that reversible hydrogenase plays a role in dark anaerobic degradation of carbon reserves with hydrogen being produced as an electron sink [35, 79].

It has been reported that the unicellular cyanobacterium *Cyanothece* 7822, is capable of hydrogenase-catalyzed hydrogen production in vivo under anaerobic conditions in the dark without the addition of an artificial reductant such as methyl viologen [80]. This study showed that it may possess the capacity to degrade carbon reserves, resulting in the evolution of hydrogen at a rate of 2.7 μmol mg^{-1} Chl per h. Asada and co-workers [35] showed that *Spirulina platensis* and *Microcystis aeruginosa* produce hydrogen in hydrogenase-mediated reactions under dark and anaerobic or microaerobic conditions. In *A. variabilis* [81] and the unicellular *Chroococcidiopsis thermalis* [82] hydrogenase is reversibly inhibited by light. The presence of light or oxygen resulted in consumption of evolved hydrogen in both *S. platensis* and *M. aeruginosa* [35]. Photobleached cells of *A. variabilis* are capable of hydrogenase-catalysed hydrogen production [83].

In the light hydrogenase-mediated hydrogen production occurs in the nonheterocystous filamentous cyanobacterium, *Oscillatoria limnetica*. However when *O. limnetica* is incubated in the presence of sulphide, photosynthetic oxygen evolution is inhibited and adaptive changes occur. This allows transfer of electrons from sulphide to Photosystem I-dependent reactions including hydrogen evolution [84].

The requirement for illumination during growth in order to exhibit hydrogenase activity probably reflects an energy requirement for cell metabolism (protein synthesis).

2.4 Factors Affecting Hydrogen Production

Various factors may have an influence on cyanobacterial hydrogen production: light intensity, gas atmosphere, temperature, composition of the growth medium, immobilization, etc. A knowledge of those factors can help to optimize cyanobacterial hydrogen production.

2.4.1 Species and Strains

Rates of H$_2$ production can vary greatly in different species. Screening of cyanobacteria from different ecosystems may provide suitable H$_2$ producers. Thus far more research has been undertaken with heterocystous cyanobacteria and it was noted, however, that the nonheterocystous marine cyanobacterium *Oscillatoria* sp. shows higher rates of hydrogen photoproduction than heterocystous *A. cylindrica* [85].

2.4.2 Light and Dark Conditions

Light is an essential factor for hydrogen evolution by cyanobacteria since hydrogen evolution depends directly or indirectly on the rates of photosynthetic reactions. Hydrogen production usually increases with increasing light intensity [86]. However, at high light intensities hydrogen production is associated with high oxygen production rates and is rapidly inhibited [87]. Outdoor conditions of full midday sunlight intensities [i.e. 2000 $\mu E\, m^{-2}\, s^{-1}$] were found inhibitory to hydrogen production, particularly after many days of exposure [77].

On the other hand, short-term illumination of cyanobacterial cells with a very high light intensity 100 W m^{-2} resulted in a rapid suppression of oxygen evolution. Photoinhibition of oxygen evolution was accompanied by stimulation of nitrogenase activity and hydrogen production [88, 89].

The relationship between hydrogen production and light intensity is dependent on the culture age, gas phase and density of culture [89]. At later stages of growth, the efficiency of light conversion to hydrogen production decreased.

Hydrogen evolution also depends on light quality [90].

Certain cyanobacteria produce hydrogen in the dark in the presence of exogenous carbohydrates, utilizing energy produced in respiration related with oxygen uptake [91]. The nitrogenase activity of *A. variabilis* and *A. cylindrica* in the dark depends on the oxygen concentration with a maximum being observed at 30% oxygen in the gas phase [92].

The rates of hydrogen production by cyanobacteria in the dark are much lower than in the light. *C. fritschii*, when grown heterotrophically in the dark with sucrose, produced hydrogen at a rate of 30–40 nmol $mg^{-1}\, h^{-1}$ during 2.5 h of incubation under an argon atmosphere [88].

There are some indications that the dark-light illumination can increase hydrogen production compared to continuous illumination in cyanobacteria [93].

2.4.3 Age of Cyanobacterial Culture and Cell Density

The hydrogen production rate depends on the age of the culture with the maximum rate of hydrogen photoproduction being observed at the beginning of the stationary phase [32, 94]. H_2 production decreased in older cultures [31]. In contrast, the oxygen-evolving capacity and photosynthetic pigment content decreased steadily with time [95]. There are not many data on the influence of cell density on cyanobacterial H_2 production. *Anabaena* sp. TU37-1 produced 0.88 ml H_2 per ml cell suspension during 24 h incubation when the cell density was 30 μg chlorophyll per ml cell suspension and light intensity 150 $\mu E\, m^{-2}\, s^{-1}$ [96]. Higher yields of hydrogen were obtained from dense than from non dense suspensions [95].

2.4.4 Temperature and pH

The optimum temperature for hydrogen production varied considerably with the organism. In case of *C. fritschii*, growth, nitrogenase activity and hydrogen evolution are optimum at 35 °C [15, 88]. The effect of temperature on the rate of hydrogen production by *Oscillatoria* sp. strain Miami BG7 varied depending on light intensity [95]. The optimum temperature for hydrogen production by *Oscillatoria* sp. strain Miami BG7 was 40 °C under a light intensity of 90 $\mu E\ m^{-2}s^{-1}$ and 30 °C under 5 $\mu E\ m^{-2}s^{-1}$. Temperature conditions that are optimal for growth of cyanobacteria may not necessarily be optimal for H_2 production [31].

Hydrogen photoproduction did not occur at pH values below 6.5 or above 10 in cyanobacteria [88]. The decrease in hydrogen photoproduction was much more pronounced at acidic than at alkaline pH. A high pH (more than 9) is unfavourable for photoproduction of hydrogen because an active uptake hydrogenase functions optimally at this pH.

2.4.5 Culture Medium

2.4.5.1 CO$_2$

Like all phototrophic organisms, cyanobacteria use carbon dioxide for photosynthesis. Cyanobacterial cultures grown under limiting CO_2 conditions have hydrogen production rates proportional to their growth rates [97]. The maximum hydrogen production rate by *A. cylindrica* was observed under 3% CO_2 in a gas phase balanced by argon [98]. Rates of hydrogen production during the incubation of cyanobacterial cells in an argon atmosphere depend on the concentration of CO_2 under which the cyanobacteria were grown [98].

In nonheterocystous cyanobacteria, CO_2 inhibits nitrogenase probably by competing for ATP and reductant [42].

2.4.5.2 N$_2$

Molecular nitrogen, which is the substrate for nitrogenase, inhibits nitrogenase-catalysed hydrogen production in some cyanobacteria [99]. About 15% N_2 in the gas phase completely inhibits hydrogen photoproduction. However, small amounts of N_2 (up to 1%) can increase hydrogen production [100]. The inhibitory effect of nitrogen on hydrogen production by *A. cylindrica* is relieved by low concentrations of carbon monoxide (an inhibitor of all nitrogenase reactions except the hydrogen-producing reaction of nitrogenase) and acetylene (an inhibitor of hydrogenase) [101]. The presence of N_2 in the growth medium is essential for long-term hydrogen production since it is necessary for nitrogen fixation and thus ultimately for cell metabolism. However, when nitrogenase catalyzes H_2 evolution only, it does not fix nitrogen [6]. Under a low pressure of nitrogen gas (e.g. partial vacuum or computer control the nitrogen supply) long-term hydrogen production is possible.

2.4.5.3 O_2

Nitrogenase as well as reversible hydrogenase are oxygen labile enzymes. However, cultures of the heterocystous *A. cylindrica* exhibit no significant oxygen inhibition of nitrogenase activity at atmospheric oxygen tensions [98]. When cultures of *A. cylindrica* were incubated in the light under a range of O_2 tensions, the highest rates of hydrogen production were obtained with 1–2% O_2 in an argon gas phase [102]. Low O_2 tensions of 1 and 2% did not inhibit H_2 evolution even after 23 h of incubation of *A. cylindrica*, although higher O_2 tensions of 5 and 10% inhibited H_2 formation by more than 50% after 23 h of incubation. In contrast, O_2 at concentrations above 1% completely inhibited hydrogen production in the nonheterocystous *Oscillatoria* sp. Miami [95].

2.4.5.4 Fixed Nitrogen (Nitrate, Ammonium, etc.)

Nitrogenase catalyzed hydrogen evolution is inhibited by the presence of fixed nitrogen (ammonium, NO_3, NO_2 and urea) in the growth medium. Combined nitrogen inhibits nitrogenase synthesis and differentiation of heterocysts from vegetative cells.

Ammonium results in inhibition of nitrogenase activity in cyanobacteria [100, 103]. However, the effect of ammonium is reversible, and when this compound is assimilated by the cells, nitrogenase becomes active again.

2.4.5.5 Physiologically Active Compounds and Carbohydrates

Photoproduction of hydrogen in *A. variabilis* was stimulated up to 7-fold by the addition of a cell extract of the water fern, *Azolla caroliniana* to the medium [8, 104]. When chloramphenicol, an inhibitor of protein synthesis, was added to *Anabaena* cells in the presence of the *Azolla* extract, the rate of hydrogen photoproduction did not increase. This indicates that physiologically active compounds in the *Azolla* extract induce additional synthesis of nitrogenase or other protein components of the nitrogen fixing system.

Nitrogenase activity and hydrogen photoproduction by cyanobacteria can be enhanced in the presence of exogenous carbohydrates [88, 91]. Fructose stimulated nitrogenase activity in *A. azollae* [105]. It has been proposed that exogenous carbohydrates protect the nitrogenase from oxygen [69].

2.4.5.6 Metal ions. Vanadium. Sulphide.

Hydrogenase activity is stimulated by the divalent ions Zn^{2+}, Ni^{2+}, Mn^{2+}, Mg^{2+}, Co^{2+} and Fe^{2+} [106]. Nickel is involved in several biological processes and low concentrations are required for the synthesis of active hydrogenase. The presence of nickel in the culture medium reduces the net hydrogen production by the enhancement of hydrogen uptake activity [107, 108]. H_2 evolution by cyanobacteria depends on the supply of growing cultures with iron [93]. An iron concentration of 5 mg l^{-1} Fe^{3+} in the medium is sufficient for active hydrogen production.

Hydrogen photoproduction in *A. variabilis* catalyzed by vanadium nitrogenase was 4 times higher than hydrogen photoproduction catalyzed by molybdenum nitrogenase [8, 109]. In order to express the alternative nitrogenase, cyanobacterial cells were grown in the medium without MoO_4^{2-} but with VO_3^- (95 nM) [109].

After 48 h in the presence of sulphide at concentrations $> 4 \text{ mM}$ the anoxygenic mode of photosynthesis was induced in *Oscillatoria limnetica* [84, 110]. In this case sulphide provided the reductant for H_2 evolution catalyzed by the reversible hydrogenase, but the cells exhibited only about 15% of the maximal hydrogenase activity in case of methyl viologen-dependent H_2 evolution [42].

2.4.6 Molecular Hydrogen

Molecular hydrogen in high concentrations (up to the 50% in the gas phase) inhibits nitrogenase activity and photosynthesis in cyanobacteria [111]. This can lead to the inhibition of hydrogen photoproduction as well. Oxygen evolution by *Anabaena variabilis* is inhibited by 40–50% after 2 h exposure to an atmosphere containing hydrogen in the light [112].

2.4.7 H_2 Uptake

Most heterocystous and some nonheterocystous cyanobacteria possess an active H_2 uptake system. Maximization of net hydrogen production by some heterocystous cyanobacteria includes minimization of hydrogen consumption catalyzed by the so-called uptake hydrogenase or/and by reversible hydrogenase.

H_2 consumption in the dark depends on O_2 uptake according to the equation 4:

$$H_2 + 0.5O_2 = H_2O \text{ (oxyhydrogen reaction)} \tag{4}$$

Uptake hydrogenase and to a small extent reversible hydrogenase catalyze this reaction (Fig. 1; Table 1) [42, 113]. Uptake hydrogenase is situated mainly in heterocysts and to a small extent in vegetative cells [114].

The rate of dark uptake of H_2 depends on the concentration of O_2. The maximum H_2 uptake by *A. cylindrica*, *A. variabilis* and *Anacystis nidulans* was observed under a concentration of O_2 of 10–15 mM in cell suspension [41, 45]. The concentration of H_2 is also important; the maximum of H_2 uptake by *A. variabilis* occurred in the presence of 10 mM H_2 [115].

Table 1. Comparative properties of reversible and uptake hydrogenases of cyanobacteria [42, 125]

	Reversible hydrogenase	Uptake hydrogenase
Localization	Cytoplasmic membrane	Thylakoid membrane
H_2 production	Catalyzes H_2 production in dark under physiological conditions as well as with artificial donors	Catalyzes H_2 production only with artificial donor such as dithionite + methyl viologen
Stability to atmospheric O_2 levels	Irreversible inactivation	Stable
Stability to 70 °C heat treatment	Stable	Not stable

In the light, uptake hydrogenase may catalyse the consumption of hydrogen and transfer electrons to the photosynthetic electron transport chain in the heterocysts. This reaction is Photosystem I-dependent, although it is not clear whether light is required for ATP synthesis, electron transport, or both [116]. The most likely initial electron acceptor for uptake hydrogenase is plastoquinone or b-type cytochrome, both in the dark and in the light. Plastoquinone appears to participate in both the photosynthetic and respiratory chains in cyanobacteria [53, 113]. The Photosystem I light-dependent reaction of uptake hydrogenase occurs only under anaerobic conditions with the inhibition of Photosystem II activity or under illumination by far red light (> 675 nm) [52, 116]. Under aerobic conditions, cyanobacteria can take up H_2 only under carbohydrate starvation.

Donation of electrons from H_2 to nitrogenase in cyanobacteria was demonstrated by Bothe et al. [45].

Thirdly there are several ways to inhibit uptake hydrogenase activity in cyanobacteria. Since hydrogen uptake occurs mostly during the oxyhydrogen reaction, removal or utilization of oxygen is a method of halting hydrogen uptake [31]. Secondly, cyanobacteria might be genetically modified by deletion of the uptake hydrogenase gene. CO inhibits hydrogen-catalyzed uptake reactions in *Anabaena* 7120 heterocysts [117].

2.4.8 Immobilization of Cells

Cyanobacteria, when immobilized in matrices such as calcium alginate, agar, cotton, polyurethane or polyvinyl foams, hollow fibres or glass beads produce hydrogen for weeks and months [5, 9, 88, 118]. Little is known about the mechanisms which induce changes in hydrogen production when cells are immobilized. Many cyanobacteria exist naturally in an immobilized-like state, either on a surface of soil particles or in symbiosis with other organisms [119].

In comparison with batch or continuous culture in reactors where free-living cells are used for H_2 production, immobilization of cells for hydrogen

Table 2. Hydrogen photoproduction by immobilized cyanobacteria on different matrices

Microorganism, immobilization matrices	Duration	Ref.
Anabaena cylindrica, glass beads	3 weeks	[101]
Anabaena azollae, polyvinyl foams	1 week	[13]
Oscillatoria sp., Miami BG 7, agar	3 weeks	[120]
Lyngbya sp., calcium alginate gel	8 days	[182]
Anabaena azollae, polyvinyl or polyurethane foams	5 months	[123]
Anabaena variabilis, hollow fibres	5 months	[9]

production may offer certain advantages such as:

– Stability of hydrogen production for several weeks or months (Table 2) [9, 101, 120, 121].

– Increased hydrogen production rates [59, 102, 120, 122, 123].

– Higher tolerance to exogenous oxygen and to high light intensity under outdoor conditions [7].

– Increasing the frequency of heterocysts and of the activity of nitrogenase [61].

– Increasing photosynthetic activity (pigment content, photosystems activity, etc.) and cells metabolism [5, 118].

– Increasing surface-to-volume ratio, which maximizes contact between the immobilized cells, growth medium and gas phase and results in accelerated growth of the cells [88].

Lambert et al. [102] reported that *A. cylindrica* immobilized on glass beads catalyzes hydrogen production at 30-fold greater rates than free-living cells. An outdoor system using *A. cylindrica* immobilized in small glass beads produced hydrogen for over three weeks [101]. The rates of hydrogen production by immobilized cells of *Oscillatoria* were a third higher than with free-living cells [120].

Hall et al. [123] immobilized *A. azollae* in polyurethane or polyvinyl foams. After 5 months of immobilization hydrogenase-catalyzed hydrogen production by the immobilized cyanobacterium was double that of free-living cells. A laboratory scale photobioreactor for photoproduction of hydrogen by immobilized *A. variabilis* on hollow fibres has recently been assembled. Photoproduction of hydrogen was measured under partial vacuum, which is cheaper and more convenient for practical bioreactors than using inert gases such as argon. Continuous hydrogen photoproduction was observed in this photobioreactor for more than 5 months [9].

3 Biochemistry and Genetics of Hydrogen Production by Cyanobacteria

3.1 Biochemistry

Hydrogen production by cyanobacteria is catalyzed by the action of either nitrogenase or the so-called reversible hydrogenase.

3.1.1 Nitrogenase

It is now accepted that there are probably three distinct nitrogenases in cyanobacteria and other organisms: MoFe nitrogenase, a vanadium nitrogenase and iron nitrogenase (or third nitrogenase) that apparently is devoid of both molybdenum and vanadium [125]. Cells of cyanobacteria grown with vanadium and without molybdenum, form vanadium-containing nitrogenase [126].

MoFe nitrogenase is an enzyme system of two distinct proteins. Dinitrogenase (MoFe protein) is a protein of molecular mass 220–240 kDa, that binds and reduces substrates. Dinitrogenase reductase (Fe protein) has the role of passing electrons to dinitrogenase and binds ATP molecules (molecular mass 51–73 kDa). Recently the latest crystallographic structure of nitrogenase Fe protein and structural models for metal centres in MoFe protein in the bacterium *Azotobacter vinelandii* have been published [127, 128].

At 30 °C in the presence of saturating N_2, FeMo nitrogenase catalyzes Eq. (5):

$$8H^+ + 8e^- + N_2 = 2NH_3 + H_2 \tag{5}$$

Each electron transfer is accompanied by the hydrolysis of two ATP molecules. About 40% of total cellular ATP is necessary for nitrogenase activity. Mg ions are necessary as well.

Hydrogen production requires that the nitrogenase reaction must utilize over 25–30% of its electrons in reducing protons to hydrogen [129]. In the case of Mo-independent nitrogenases the proportion of the electron flux resulting in hydrogen evolution in the presence of N_2 is significantly higher than for Mo nitrogenase. For vanadium nitrogenase it is 50%, and for the third nitrogenase it is 75% of the electron flux being utilized in this way [130].

In the absence of N_2, nitrogenase reduces protons to hydrogen.

3.1.2 Hydrogenase

Hydrogenase is a protein which catalyzes the reversible activation of hydrogen according to Eq. (6):

$$H_2 \rightleftharpoons 2H^+ + 2e^- \tag{6}$$

Unlike nitrogenase, cyanobacterial hydrogenase is different from the hydrogenases of other microorganisms.

There are at least two distinct groups of hydrogenase proteins: the enzymes containing Fe as the only metal have several Fe–S clusters, while the Ni–Fe hydrogenases contain at least one Fe–S cluster and a Ni atom which is covalently attached to the protein. There is now agreement among most investigators that the Ni–Fe-hydrogenases are composed of two subunits of molecular weights of around 60 000 and 25 000 [131].

Cyanobacteria possess two different hydrogenases. One enzyme, the so-called uptake hydrogenase, is found in the membrane fraction when cyanobacterial cells are disrupted; it is involved in the consumption of hydrogen. The other enzyme, which catalyzes both the uptake and the evolution of hydrogen in vitro (reversible hydrogenase) is found in the supernatant when cells are disrupted and centrifuged and is therefore considered to be a soluble enzyme. Recently, however, it was shown that this "soluble" enzyme is a component of the cytoplasmic membrane in *Anacystis nidulans* [125].

Reversible hydrogenase can catalyze hydrogen production with artificial electron donors such as methyl or benzyl viologen, chemically reduced by dithionite [42]. Among the physiological electron donors, the simultaneous presence of only pyruvate and CoA resulted in hydrogen evolution at a similar rate to dithionite [35]. The enzyme is sensitive to oxygen; oxygen concentrations as low as 0.1% inhibit hydrogenase activity and thus activity is usually demonstrated under anaerobic conditions [41]. Activation of the enzyme requires, anaerobic preincubation in the dark for several hours [42].

Reversible hydrogenase is very stable to heat without loss of activity after 1 h at 70 °C [42].

3.2 Genetics

The genetics of cyanobacterial hydrogen production has received little investigation since most attention has been on the role of nitrogenase in nitrogen fixation.

Genetic methods for the study of cyanobacteria are discussed by Haselkorn [132]. They include methods for induction and selection of mutants, methods for introduction of DNA into cells, and methods for selection and analysis of complemented mutants and recombinants.

The possible objectives of genetic work related to cyanobacterial hydrogen photoproduction include:

– Investigation of genes controlling the proportion of cells that differentiate to heterocysts [133].

– Investigation of hydrogenase genes aimed at deletion of "uptake" hydrogenase activity.

- Optimization of photosynthetic conversion efficiency for hydrogen production.
- Obtaining mutants defective in alternative electron "sinks" than hydrogen.

3.2.1 Nitrogenase Genes (nif)

Both nitrogenase and hydrogenase are complex enzymes whose synthesis requires the action of a large number of accessory genes and whose expression is regulated by the products of several regulatory genes. In addition, the three nitrogenase systems (Mo, V, Fe) are genetically distinct, being encoded by different structural genes [130].

Nitrogenase genes can be divided into three categories according to the classification presented above and describe the relation between cyanobacterial nitrogenase and molecular oxygen. The most detailed study of nif gene organization in cyanobacteria has been carried out with heterocystous Anabaena [132, 134–136].

A number of genes that are turned on or off in Anabaena heterocyst differentiation have been cloned and sequenced. Most attention has been focussed on the three genes nifH, nifD and nifK, which encode the polypeptide components of the nitrogenase complex. The nifH gene codes for the structural unit of dinitrogenase reductase, and nifD and nifK for the structural units of dinitrogenase. In contrast to other microorganisms, in the vegetative cells of cyanobacteria a large segment of DNA separates the nifK gene from nifHD genes. It seems that the DNA separating nifK and nifHD does not contain nif structural genes [138]. During heterocyst differentiation, this DNA segment is removed and nitrogenase activity initiated [138]. Results of a study by Elhai and Wolk [136] support the view that the induction of nif genes is regulated by a developmental signal, associated with the process of heterocyst differentiation. Alternatively, the induction of nif genes in vegetative cells may be under environmental control, i.e. the absence of fixed nitrogen (ammonia) or oxygen. It is not clear yet how this regulation occurs.

A different situation occurs in nonheterocystous filamentous and unicellular cyanobacteria. These cyanobacteria have a contiguous nifHDK operon. The marine cyanobacterium, Trichodesmium fixes nitrogen aerobically in light and when grown under a light-dark cycle fixes nitrogen only during the light phase. It was found that no gene rearrangements occur when the cultures were grown with combined nitrogen or under nitrogen-fixing conditions [139].

Cyanobacteria like Plectonema, which show nitrogenase activity only under anaerobic conditions, probably have a contiguous nifHDK operon [140, 141].

3.2.2 Hydrogenase Genes

Genetic investigations of hydrogenase have only just begun. Recently the nucleotide sequence of the gene proposed to encode the small subunit of the reversible hydrogenase of the thermophilic unicellular *Synechococcus* PCC 6716 [142] and the heterocystous *A. cylindrica* [107, 109] has been isolated.

A major aim of genetic work with "uptake" hydrogenase is to produce hydrogenase-deficient (*hup⁻*) strains of cyanobacteria. Hybridization DNA from *A. cylindrica* and three plasmids containing cloned hydrogenase genes from the bacterium *Bradyrhizobium japonicum* have been made [107].

4 Photobioreactor Development

Only a few reports are available on cyanobacterial photobioreactors for hydrogen photoproduction (Table 3) [9, 76, 101, 143–145].

An outdoor photobioreactor system in Canberra (Australia) using *A. cylindrica* B629 suspended in glass beads was shown to produce H_2 continuously for over three weeks with total hydrogen produced up to 1.2 l [101]. For this photobioreactor a gas-tight glass box with outside dimensions $136.8 \times 108.6 \times 13$ cm was used. The gas phase of the box (158 litre) was CO (0.2%), C_2H_2 (5%), O_2 and N_2. C_2H_2 is an inhibitor of hydrogenase. 6.2 g (dry weight) of cyanobacterial cells were added to the photobioreactor. The temperature of the box was maintained below 30 °C at all times by means of the water reservoir on top of the lid. The photobioreactor vessel was covered by a shadecloth which reduced the natural illumination by approx. 70%. The illumination was different from day to day. The pH at the beginning of the experiment was 7.0 and at the end was 11.

A series of 5-litre outdoor photobioreactors with *Oscillatoria* sp. produced hydrogen for over a week at rates of 8.2 ml $g^{-1} h^{-1}$ in Florida (USA) [144]. The photobioreactor was designed in the general form of solar panels to permit efficient use of incident sunlight. The light intensity was adjusted by using neutral density material. Full midday sunlight intensities (i.e. 2000 μE $m^{-2} s^{-1}$) were found to be inhibitory to H_2 production. Cells for these experiments were cultured on a limited amount of ammonia as the nitrogen nutrient source. Optimal rates of H_2 production were observed after cultures became depleted of

Table 3. Hydrogen photoproduction by cyanobacteria in photobioreactors

Microorganism	Rate of H_2 production (ml per g chloroph. per h)	Duration	Ref.
Oscillatoria sp.	8.2	1 week	[144]
Anabaena azollae	up to 40	6 days	[13]
Anabaena cylindrica	—	3 weeks	[101]
Phormidium valderianum	0.4	25 days	[76]
Anabaena variabilis	200	5 months	[9]
Anabaena cylindrica	—	5 weeks	[143]

nitrogen nutrients. The temperature of the cyanobacterial culture was maintained constant at $26 \mp 4\,°C$.

Park et al. [145] used a column photobioreactor (volume 64 ml) with polyurethane immobilized *A. azollae* for H_2. The photobioreactor column was filled with 100 pieces of sterilized polyurethane foam (5-mm^3). The medium $(6.25\ ml\ h^{-1})$ was trickled through the foams. Illumination was with fluorescent lamps $(100\ \mu E\ m^{-2}\ s^{-1})$ and the temperature was 27 °C. Total H_2 production reached 2 ml H_2 per reactor (total 4 mg chloroph.) after 6 d under anaerobic conditions. Hydrogen production was measured with the presence of CO (an inhibitor of all nitrogenase activity except H_2 production).

Cyanobacterial hydrogen production in the above cases was operated as a two-step process: (1) a long-term growth phase for the production of hydrogen-generating biomass and (2) a short-term hydrogen production phase using stored carbon reserves under anaerobic conditions. The photobioreactor can be also operated as a continuous hydrogen photoproduction [9]. In this case column continuous-flow photobioreactors (volume 15 and 70 ml) were used with hollow-fibre immobilized cyanobacteria (Fig. 2). The photobioreactor was designed so that the cyanobacterial growth medium passes from the outside of the fibres into the inner lumen space. Cyanobacterial cells weighing 70 mg (dry weight) were added to the photobioreactor. The photobioreactor column was maintained at 28 °C and illuminated continuously with a fluorescent lamp $(25\ \mu E\ m^{-2}\ s^{-1})$. The flow rate of the medium was 8 ml h^{-1}. Photoproduction of

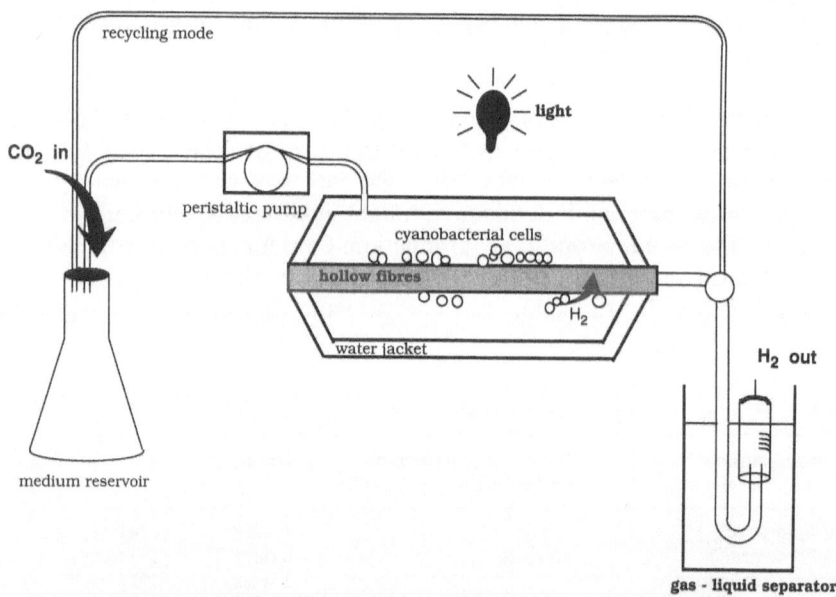

Fig. 2. Schematic diagram of hollow fibre photobioreactor for continuous production of hydrogen by immobilized cyanobacteria

hydrogen at rates up to $200 \, \text{ml} \, \text{g}^{-1} \, \text{h}^{-1}$ in a hollow fibre photobioreactor with immobilized *A. variabilis* was observed for more than 5 months.

More studies are available of the use of photobioreactors: in general [146–148] for biomass production with algae [10, 11, 149, 150], CO_2 removal [151, 152], ammonium photoproduction [153, 154], removal of nitrate from water [12, 155], photoproduction of hydrogen peroxide [145], photoproduction of anthocyanin [156] and removal of organochlorine compounds [157]. Several reports are available about computer-controlled photobioreactors [10, 11, 158]. Pirt and co-workers [10] used computer control to maintain algal biomass concentrations in a tubular photobioreactor. Walach et al. [11] used computers to control the ratio of carbon to nitrogen in a photobioreactor containing *Spirulina platensis*. The computer-controlled systems maintained the algal biomass at a constant concentration while fluctuations in CO_2 partial pressure and light intensity occurred [158].

A few reports are available on hydrogen-producing photobioreactors on the laboratory scale using purple bacteria [159–167] and *Halobacterium halobium* [168, 169]. In contrast to cyanobacteria where water is the substrate for hydrogen production, purple bacteria require organic compounds to be oxidized for hydrogen production.

The hydrogen production photobioreactor usually includes:
1) medium reservoir
2) light source
3) pump for circulating the medium
4) column or tank with cyanobacteria or other microorganisms
5) H_2 collector.

There are some important considerations for optimum design of a hydrogen production photobioreactor:

– The photobioreactor column or tank must be gas tight to avoid the diffusion of hydrogen. On the other hand there must be continuous flow of the medium through photobioreactor to avoid hydrogen consumption by the cyanobacterial cells [9].

– Photobioreactors are usually illuminated from outside the reaction vessel, from one or several sides [9, 124, 144] or internally using optical fibres as light guides [164]. The use of a bundle of optical fibres which diffuse the light out through the surface perpendicular to the axis of the fibre, allows good light distribution throughout the culture [170]. In the case of external illumination, the geometry of the photobioreactor is calculated for the most effective utilization of light energy. Lee and Low [150] showed that the overall algal biomass output rate of a tubular-loop photobioreactor can be altered by inclining the bioreactor at an angle to the horizontal. Indoor photobioreactors usually use fluorescent lamps with a spectrum somewhat similar to solar irradiation [9, 13, 167].

– Hydrogen production in a photobioreactor is induced by microaerobic conditions (atmosphere with low N_2 and O_2 partial pressure). Anaerobic conditions are created in a photobioreactor by flushing with an inert gas (Ar or He) [101], under partial vacuum [9] or using a liquid paraffin layer [162].

– The temperature of the photobioreactors is controlled by thermostatted water flow [9, 163] or by placing the bioreactor into a thermostatted water bath [160].

– In order to control the hydrogen production rate it is essential to grow the cyanobacteria free of external contaminations such as other algae, bacteria, fungi and zooplankton. This can be achieved by employing closed cultivation systems, designed for the axenic mass cultivation [171].

– Both free [85, 162] and immobilized cells [5, 9, 101, 145, 164] can be used in the photobioreactors for hydrogen production.

The development of hollow fibre immobilization techniques for continuous flow photobioreactors appears promising [9, 172–175]. In this case the system is built around bundles of hollow fibres composed of semipermeable polymeric membranes. Hollow fibres have been successfully used in biotechnology as filter aids and as supports for immobilization of enzymes, microbial, animal and plants cells [176, 177].

A novel hollow fibre photobioreactor has been designed for continuous hydrogen production using hollow fibre immobilized cyanobacteria [9]. Photoproduction of hydrogen at rates up to 200 ml $g^{-1} h^{-1}$ was observed for more than 5 months.

Hollow fibre photobioreactors have several advantages.

– The large surface area-volume ratio of the hollow fibres allows the design of a compact system.

– Separation of H_2 from the cell culture, enables avoidance of H_2 consumption.

– Stabilization and enhancement of the life-time of hydrogen photoproduction.

5 Conclusions

Compared to other hydrogen producing organisms (purple bacteria, green algae) cyanobacteria are unique in their ability to produce hydrogen using water as their ultimate electron source and solar energy as an energy source under air conditions. In addition they are fast-growing (doubling-times of the order of 20 h) with a flexible metabolism.

Research progress over the last decades toward scaled-up cyanobacterial hydrogen photoproduction appears promising.

– Cyanobacterial hydrogen production systems can be operated continuously for a considerable time, 5 months or more [9].

– Cyanobacterial hydrogen production systems can be operated at good efficiencies in outdoor systems [85, 101].

– There are certain metabolically unique strains of cyanobacteria which have special promise for photobioreactors for hydrogen production [144].

– It is possible to manipulate hydrogen metabolism genetically in cyanobacteria [107, 153, 178].

– It is possible to manipulate hydrogen production biochemically in cyanobacteria. In the case of Mo-independent nitrogenase, hydrogen production is significantly higher than for Mo-containing nitrogenase [6, 8, 179].

– There are sufficient data on physiological conditions of hydrogen production (light intensity, gas atmosphere, temperature, immobilization, influence of physiologically active compounds, etc.) which can assist in optimizing H_2 photoproduction systems [5, 8, 91, 98, 106].

Immobilization of cyanobacteria in some substrates such as agar, polyurethane or polyvinyl foams, hollow fibres, glass beads and cotton led to an increase and/or stabilization of the rate of H_2 photoproduction. Photoproduction of hydrogen at high rates (up to $200 \, ml \, g^{-1} h^{-1}$) was observed in a photobioreactor with hollow fibre immobilized cells *A. variabilis*.

– Simultaneously with H_2 evolution cyanobacteria consume CO_2 and may thus provide a mechanism for CO_2 removal from air [152]. Some cyanobacterial hydrogen producers are also excellent feed material for aquaculture, cattle, etc. [85, 29, 181].

Despite progress in photobioreactor development, there are some problems to be solved. These include:

– Separation of H_2 and O_2 produced.

– Supply of CO_2 and N_2 for the cyanobacterial growth.

– Outdoor high temperatures conditions and high light intensities which can inhibit growth.

With design of computer-controlled photobioreactors many of these problems can be solved. Several parameters could be computer controllable in a photobioreactor such as gas supply (CO_2, N_2 and perhaps O_2), pH, rates of liquid flow, etc. Some potentially uncontrollable variables such as temperature and light intensity must be carefully considered as well. Computer control can allow maximum yields of H_2 production with minimal O_2 and CO_2 evolution.

Acknowledgements. Authors thank the New Energy and Industrial Technology Development Organization and Research Institute of Innovative Technology for the Earth (Japan) for financial support.

6 References

1. Bolton JR, Hall DO (1979) Ann Rev Energy 42: 353
2. Harriman A, West MA (1982) Photoproduction of hydrogen. Academic, London
3. Grätzel M (ed) (1983) Energy resources through photochemistry and catalysis. Academic, New York
4. Veziroglu TN, Takahashi PK (eds) (1990) Hydrogen Energy Progress VIII. Proceedings of the 8th World Hydrogen Energy Conference. Honolulu and Waikoloa, Hawaii, USA. 22–27 July 1990. (in Three Volumes) Pergamon, New York, p 1594
5. Hall DO, Rao KK (1989) Chimicaoggi 7: 40
6. Bothe H, Kentemich T (1990) Potentialities of H_2 production by cyanobacteria for solar energy conversion programs. In: Veziroglu TN, Takahashi PK (eds) Hydrogen Energy Progress VIII. Proc. 8th World Hydrogen Energy Conference, Honolulu and Waikoloa, Hawaii, USA, 22–27 July 1990. Pergamon, New York, p 729
7. Mitsui A, Kumazawa, S, Takahashi A, Ikemoto H, Suda S, Hanagata N, Domeier M, Komatsu M, Benkert J (1990) Overview on the biological hydrogen photoproduction research. In: Hydrogen Photoproduction Workshop IV. Waikoloa, Hawaii, 26–27 July 1990, p 171
8. Gogotov IN, Troshina OY (1990) Biotechnological foundations for hydrogen production by employing phototrophic microorganisms. In: Hydrogen Photoproduction Workshop IV. Waikoloa, Hawaii, 26–27 July 1990, p 123
9. Markov SA, Rao KK, Hall DO (1992) A hollow fibre photobioreactor for continuous production of hydrogen by immobilized cyanobacteria under partial vacuum. In: Veziroglu TN, Derive C, Pottier J (eds) Hydrogen energy progress IX. Proc. 9th World Hydrogen Energy Conference, Paris 1992, MCI, Paris, p 641
10. Pirt SJ, Lee YK, Walach MR, Pirt MW, Balyuzi HHM, Bazin M (1983) J Chem Tech Biotechnol, 33B: 35
11. Walach MR, Bazin M, Pirt JS, Balyuzi HM (1986) Biotech Bioeng 29: 520
12. Garbisu C, Gil JM, Bazin MJ, Hall DO, Serra JL (1991) J Appl Phycology 3: 221
13. Park IH, Rao KK, Hall DO (1991) Biochem. (Life Sci. Adv.) 10: 173
14. Lee YK, Low C-S (1992) Biotechnol Bioeng 40: 1119
15. Fay P (1965) J Gen Microbiol 35: 11
16. Okhi K, Katoh T (1975) Plant and Cell Physiology 16: 53
17. Smith AJ (1983) Ann Microbiol (Inst Pasteur) 134B: 93
18. Markov SA, Porshneva EB, Nikolaeva LF (1985) Stability of photosynthetic apparatus and nitrogenase system in the blue-green alga in the darkness. In: Environmental Conditions and Plant Productivity. East Siberian Book Publisher, Irkutsk, p 125 (in Russian)
19. Padan E (1979) Adv Microbial Ecol, 3: 1
20. Gusev MV (1968) The Biology of Blue-green Algae. Moscow University Publisher, Moscow (in Russian)
21. Carr NG, Whitton BA (eds) (1973) The Biology of Blue-green Algae. Blackwell, Oxford
22. Fogg GE, Stewart WDP, Fay P, Walsby AE (1982) The Blue-green Algae. Academic Press, London
23. Gusev MV and Nikitina KA (1979) Cyanobacteria: Physiology and Metabolism, Nauka, Moscow (in Russian)
24. Humm HJ, Wicks SR (1980) Introduction and guide to the marine bluegreen algae. Wiley, New York
25. Fay P (1983) The Blue Greens (Studies in Biology/Institute of Biology, No. 160), Camelot, Southampton
26. Carr NG, Whitton BA (eds) (1982) The biology of cyanobacteria. Blackwell, Oxford
27. Fay P, Van Baalen C (eds) (1987) The cyanobacteria. Elsevier, Amsterdam
28. Mann NH, Carr NG (eds) (1992) Photosynthetic prokaryotes. Plenum, New York
29. Ciferri O (1983) Microbiol Rev 47: 551
30. Shestakov SB (1984) Perspectives of using phototrophic bacteria in biotechnology. In: Baev AA (ed) Biotechnology. Nauka, Moscow, p 160 (in Russian)
31. Lambert GR, Smith GD (1981) Biol Rev 56: 589
32. Benemann JR, Weare NM (1974) Science 184: 174
33. Oshchepkov VP, Nikitina KA, Gusev MV, Krasnovsky AA (1974) Doklady Akademii Nauk SSSR 213: 557
34. Bishop NI, Frick M, Jones LM (1978) Alternative fates of the photochemical reducing power

generated in photosynthesis: Hydrogen production and nitrogen fixation. In: Current Topics in Bioenergetics 8: 3
35. Asada Y, Miyake M, Tomizuka N (1992) Hydrogenase-mediated hydrogen metabolism in some cyanobacteria. In: Murata N (ed) Research in Photosynthesis. Kluwer, Dordrecht vol I, p 477
36. Fay P (1992) Microbiol Rev 56: 340
37. Adams DC, Carr NG (1981) Crit Rev Microbiol 9: 45
38. Fay P (1980) Nitrogen fixation in heterocysts. In: Subba Rao NS (ed) Recent Advances in Biological Nitrogen Fixation. Edward Arnold, London, p 121
39. Haselkorn R (1978) Ann Rev Plant Physiol 29: 319
40. Wolk CP (1982) Heterocysts. In: Carr NG, Whitton BA (eds) The Biology of Cyanobacteria. Blackwell, Oxford, p 359
41. Kondratieva EN, Gogotov IN (1981) Molecular Hydrogen in Microbial Metabolism. Nayka, Moscow (in Russian)
42. Houchins J (1984) Biochim Biophys Acta 768: 227
43. Kangatharalingam N, Dodds WK, Priscu JC, Paerl HW (1991) J Phycol 27: 680
44. Jüttner F (1983) J Bacteriol 155: 628
45. Bothe H, Distler E, Eisbrenner G (1978) Biochimie 60: 277
46. Neuer G, Bothe H (1985) Arch Microbiol 143: 185
47. Winkenbach F, Wolk CP (1973) Plant Physiol 52: 480
48. Tamagnini P, Yakunin AF, Gogotov IN, Lindblad P (1993) FEMS Microbiol Letters 107: 37
49. Bothe H (1977) Flavodoxin. In: Trebst A, Avron M (eds) Encyclopedia of Plant Physiology. New series, Photosynthesis 1. Springer, Berlin Heidelberg New York, vol 5, p 217
50. Yakunin AF, Hallenbeck PC, Troshina OY, Gogotov IN (1993) BBA (in press)
51. Benemann JR, Yovh DC, Valentine RC, Arnon DI (1969) The electron transport system in nitrogen fixation by *Azotobacter*. I. Azotoflavin as an electron carrier. In: Proc. Nat. Acad. Sci. USA 64: 1079
52. Houchins JP, Hind G (1982) Biochim Biophys Acta 682: 86
53. Eisbrenner G, Bothe H (1979) Arch Microbiol 123: 37
54. Stanier RY, Ingraham JL, Wheelis ML, Painter PR (1987) General Microbiology. Macmillan, Houndmills
55. Kerfin W, Böger P (1982): Physiol Plant 54: 93
56. Adams DG (1992) J Gen Microbiol 138: 355
57. Sahy J, Adnicary SP (1981) Zeitschr Allgem Microbiol 21: 669
58. Hill DJ (1975) Planta 122: 179
59. Brouers M, Hall DO (1986) J Biotechnology 3: 307
60. Braun-Howland EB, Nierzwicki-Bauer SA (1990) *Azolla-Anabaena* symbiosis: biochemistry, physiology, ultrastructure, and molecular biology. In: Rai AN (ed) Handbook of Symbiotic Cyanobacteria. CRC Press Boca Raton, Florida, p 65
61. Shi D-J, Hall DO (1988) Plants Today 1: 5
62. Smith RV, Evans MCW (1970) Nature 225: 1253
63. Gallon JR, Chaplin AE (1988) Recent studies on N_2 fixation by nonheterocystous cyanobacteria. In: Bothe FJ, de Bruijn FJ, Newton WE (eds) Nitrogen Fixation: Hundreds Years After. Gustav Fischer, Stuttgart, Germany, p 183
64. Gallon JR (1992) New Phytol 122: 571
65. Mullineaux PM, Chaplin AE, Gallon JR (1980) J Gen Microbiol 120: 227
66. Gallon JR, Perry SM, Rajab TMA, Flayeh KAM, Junes JS, Chaplin AE (1988) J Gen Microbiol 134: 3079
67. Rai AN, Borthakur M, Bergman B (1992) J Gen Microbiol 138: 481
68. Rippka R, Waterbury JB (1977) FEMS Microbiol Letters 2: 83
69. Nguen Thi Hoa (1985) Conditions of nitrogen fixation in cyanobacteria *Plectonema boryanum* and *Gloeothece*. Summary of PhD thesis, Moscow State University, Moscow (in Russian)
70. Suda S, Kumasawa S, Mitsui A (1992) Arch Microbiol, 158: 1
71. Kuwada Y, Nakatsukasa M, Ohta Y (1988) Agrac Biol Chem 52: 1923
72. Kuwada K, Inoue Y, Koike H, Ohta Y (1991) Agric Biol Chem 55: 299
73. Kumazawa S, Mitsui A (1985) Appl Environ Microbiol 50: 287
74. Kumazawa S, Mutsui A (1991) Plant Science Tomorrow 3: 8
75. Subramanian G, Prabaharan D (1989) Hydrogen production by marine cyanobacteria. In: Abst. The First International Marine Biotechnology Conference. Tokyo, Japan. 4–6 September 1989, p 21

76. Prabaharan D (1992) Hydrogen production by marine cyanobacteria. PhD thesis. Bharathidasan University, Tiruchrapalli, India
77. Mitsui A, Rosner D, Kumazawa S, Barciela S, Philips E, Ramachandran S, Takahashi A, Richards J (1985) Hydrogen production from salt water by marine blue-green algae and solar radiation. In: Haise FW (ed) Proceedings of Twenty-Second Space Congress. Canaveral Council of Technical Societies, p 11
78. Serebryakova LT, Zorin NA, Gogotov IN (1992) Mikrobiologia 61: 175
79. Van der Oost J, Cox RP (1989) Arch Microbiol 151: 40
80. Van der Oost J, Kanneworff WA, Krab K, Kraayenhof R (1987) FEMS Microbiology 48: 48
81. Spiller H, Bookjans G, Shanmugam KT (1983) J Bacteriol 155: 129
82. Almon H, Böger P (1988) FEMS Microbiol Lett 49: 445
83. Laczko I, Barabas K (1981) FEBS Letters 153: 312
84. Belkin S, Padan E (1978) FEBS Letters 94: 291
85. Mitsui A, Philips EJ, Kumazawa S, Reddy KJ, Ramachandran S, Matsunaga T, Haynes L, Ikemoto H (1983) Progress in research toward outdoor biological hydrogen production using solar energy, sea water, and marine photosynthetic microorganisms. In: Biochemical Engineering 111–413. Annals of the New York Academy of Sciences, p 514
86. Jeffries TW, Timourian H, Ward RL (1978) Appl Environ Microbiol 35: 704
87. Agar J, Suda S, Takeyama H, Lee W, Mitsui A (1991) Hydrogen and oxygen photoproduction by marine unicellular cyanobacterium under high light intensities equivalent to mid-day intensities. In: Abstract 2nd Int. Marine Biotechnology Conference, Baltimore, USA, 13–16 October 1991, p 75
88. Markov SA (1987) The relationship between nitrogenase activity, photosynthesis and assimilation of exogenic carbohydrates in cyanobacterium *Chlorogloeopsis fritshii*. PhD thesis, Moscow State University, Moscow (in Russian)
89. Markov SA, Polesskaya OG, Krasnovsky AA (1991) Soviet Plant Physiology 38: 652
90. Kumar D, Kumar HD (1991) Int J Hydrogen Energy 16, 397
91. Markov SA, Krasnovsky AA (1985) Soviet Plant Physiology 32: 562
92. Jensen BB, Cox RP (1983) Arch Microbiol 135: 287
93. Jeffries TW, Leach KT (1978) Appl Environ Microbiol 35: 1228
94. Peleckaya EN, Polesskaya OG, Krasnovsky AA (1986) Prikladnaya Biochimia i Mikrobiologia 22: 500 (in Russian)
95. Philips EJ, Mitsui A (1983) Appl and Environ Microbiol 45: 1212
96. Kumazawa S, Suda S, Mitsui A (1991) Effect of cell densities to the hydrogen photoproduction by marine cyanobacteria, *Synechococcus* sp. Miami BG4351 and *Anabaena* sp. TU37-1. In: Abst. 91st General Meeting of Amer. Soc. Microbiol., p 222
97. Jones BB, Bishop NI (1976) Plant Physiol 57: 659
98. Lambert GR, Smith GD (1980) Arch Biochem Biophys 205: 36
99. Lambert GR, Smith GD (1977) FEBS Lett 83: 159
100. Kondratieva EN, Gogotov IN (1983) Production of molecular hydrogen in microorganisms. In: Adv. Biochem. Engineering/Biotechnology 28: 139
101. Smith GD, Lambert GR (1981) Biotech Bioeng 23: 213
102. Lambert GR, Daday A, Smith GD (1979) Appl and Environ Microbiol 38: 521
103. Yakunin AF, Troshina OY, Jha M, Gogotov IN (1992) Mikrobiologia 61: 377
104. Troshina OY, Gogotov IV (1992) Prikladnaya Biochimia i Mikrobiologia 28: 380 (in Russian)
105. Rozen A, Arad H, Schonfeld M, Tel-or E (1981) Arch Microbiol 145: 187
106. Asada Y, Kawamura S, Ho K-K (1987) 26: 637
107. Smith GD, Ewart GD, Tucker W (1992) Int J Hydrogen Energy 17: 695
108. Tredici MR, Margheri MC, De Philippis R, Materassi R (1990) J Gen Microbiol 136: 1009
109. Yakunin AF, Chan Van Ni, IN Gogotov (1989) Dokladi Akademii Nayk SSSR 307: 1269
110. Sybesma C, Schowanek D, Slooten L, Walravens N (1986) Photosynthesis Research 9: 149
111. Antarikanonda P, Bernd H, Mayer F, Lorenzen H (1980) Arch Microbiol 69: 114
112. Polesskaya OG, Krasnovsky AA (1985) Soviet Plant Physiology 32: 79
113. Houchins JP, Burris RH (1981) Plant Physiol 68: 717
114. Lindblad P, Sellstedt A (1990) Protoplasma 159: 9
115. Eisbrenner G, Distler E, Floener L, Bothe H (1978) Arch Microbiol 118: 177
116. Benemann JR, Weare NM (1974) Arch Microbiol 101: 401
117. Peterson RB, Burris RH (1978) Arch Microbiol 116: 125
118. Shi D-J (1987) Energy metabolism and structure of the immobilized cyanobacterium *Anabaena azollae*. PhD thesis. King's College London, University of London, London

119. Egorov NS, Landaw NS, Borman EA, Kotova NB (1984) Prikladnaya Biokhimia i Mikrobiologia 20: 579 (in Russian)
120. Philips EJ, Mitsui A (1986) Int J Hydrogen Energy 11: 83
121. Rao KK, Hall DO (1992) Immobilized photosynthetic system: application in biotechnology. In: Barber J, Guerrero MG, Medrano H (eds) Trends in Photosynthetic Research. Intercept, Andover, Hampshire, p 135
122. Muallem A, Bruce D, Hall DO (1983) Biotech Letters 5: 365
123. Hall DO, Brouers M, de Jong H, De la Rosa MA, Rao KK, Shi DJ, Yang LW (1987) Photobiochem Photobiophys Suppl: 167
124. Sasikala K, Ramana CV, Rao PR (1992) Int J Hydrogen Energy 17: 23
125. Kentemich T, Casper M, Bothe H (1991) Naturwissenschaften 78: 559
126. Kentemich T, Danneberg G, Hundeshagen B, Bothe H (1988) FEMS Microbiology Letters 51: 19
127. Georgiadis MM, Komiya H, Chakrabarti P, Woo D, Kornuc JJ, Rees DC (1992) Science 257: 1653
128. Kim J, Rees DC (1992) Science 257: 1677
129. Burris RH (1991) J Biol Chem 266: 9339
130. Smith BE, Eady R (1992) Eur J Biochem 205: 1
131. Sharavankumar CH, Polasa H (1992) Indian J Microbiol 32: 29
132. Hazelkorn R (1991) Genetic systems in cyanobacteria. In: Miller JH (ed) Methods in Enzymology. Academic Press, Harcourt Brace Jovanovich, San Diego, vol 204, p 418
133. Wolk CP, Elhai J, Cai Y, Bancroff I, Panoff J-M (1990) Summary statement for hydrogen production workshop IV. In: Hydrogen Photoproduction Workshop IV. Waikoloa, Hawaii, 26-27 July 1990, p 107
134. Haselkorn R (1986) Ann Rev Microbiol 40: 525
135. Mulligan ME, Haselkorn R (1989) J Biol Chemistry 264: 19200
136. Elhai J, Wolk CP (1990) The EMBO Journal 9: 3379
137. Damerval T, Franche C, Rippka RM, Cohen-Bazire G (1985) Rearrangement of nif structural genes in Nostoc PCC7906. In: Evans HJ, Bottomley PJ, Newton WE (eds) Nitrogen Fixation Research Progress. Martinus Nijhoff, New York, p 517
138. Rice D, Mazur BJ, Haselkorn R (1982) J Biol Chem, 257: 13157
139. Zenhr JP, Ohki K, Fujita Y (1991) J Bacteriol 173: 7055
140. Kallas T, Coursin T, Rippka R (1985) Plant Mol Biol 5: 321
141. Apte SK, Thomas J (1987) J Genet 66: 101
142. Van der Oost J, Walraven HS, Bogerd J, Smit AB, Ewart GD, Smith GD (1989) Nucleic Acids Research 17: 10098
143. Miyamoto K, Hallenbeck PC and Benemann (1979) J Ferment Technol 57: 287
144. Mitsui A, Kumazawa S, Philips KJ, Reddy K, Gill K, Matsunaga T, Renuka BR, Kusumi T, Reyes-Vasquez G, Miyazawa K, Haynes L, Ikemoto H, Duerr E, Leon CB, Rosner D, Sesco R, Moffat E (1985) Mass cultivation of algae and photosynthetic bacteria: concepts and application. In: Ghose TK (ed) Biotechnology and Bioprocess Engineering. United India Press, New Delhi, p 119
145. Park IH, Rao KK, Hall DO (1991) Int J Hydrogen Energy 16: 313
146. Pulz O (1992) Cultivation techniques for microalgae in open and closed systems. In: Proc. 1st European Workshop on Microalgal Biotechnology, Potsdam, 10-12 June 1992, p 61
147. Pulz O (1993) Semiclosed tubular-Pate systems for the cultivation of microalgae aimed to health, food and pharmaceutical use. In: 6th International Conference on Applied Algology, Česke Budějovice, Czech Republic, 6-11 September 1993, p 39
148. Hall DO, Garbisu C, Kannaiyan S, Lichtl R, Markov SA, Rao KK, Serra JL, Sopko B (1993) Photobioreactors with immobilized cyanobacteria for production of fuels and chemicals and for water purification: status, potentials and problems. In: 6th International Conference on Applied Algology, Česke Budějovice, 6-11 September 1993, p 36
149. Lee ET-Y, Bazin M (1990) New Phytol 116: 331
150. Lee Y-K, Low C-S (1991) Biotechnol Bioeng 38: 995
151. Takano H, Takeyama H, Nakamura N, Sode K, Burgess JG, Manabe E, Hirano M, Matsunaga T (1992) Appl Biochem Biotechnol 34/35: 449
152. Karube I, Takeuchi T, Barnes DJ (1992) Adv Biochem Engineering/Biotechnology 46: 63
153. Vincenzini M, Brouers M, Hall DO, Materassi R (1986) Photobiochem. Photobiophys 13: 85
154. Wang SC, Jin MR, Hall DO (1991) Bioresource Technology 38: 85
155. Garbisu C (1992) Nitrate and phosphate uptake from water by free-living and immobilized

cells of the cyanobacterium *Phormidium laminosum*. PhD thesis. King's College London, University of London
156. Zhong JJ, Yoshida M, Fujiyama K, Seki T, Yoshida T (1993) J Ferment Bioeng 75: 299
157. Wu XY, Kosaric N (1991) Water Science and Technology 24: 221
158. Walach MR, Balyuzi HHM, Bazin MJ, Lee YK, Pirt SJ (1983) J Chem Tech Biotechnol 33B: 59
159. Weetall HH, Sharma BP, Detar LC (1981) Biotech Bioeng 23: 605
160. Von Felten P, Zurrer H, Bachofen R (1985) Appl Microbiol Biotechnol 23: 15
161. Hirayama O, Uya K, Hiramutsu Y, Yamada H, Moriwaki K (1986) Agric Biol Chem 50: 891
162. Kim JS, Ito K, Izaki K, Takahashi H (1987) Argic Biol Chem 51: 2591
163. Planchard A, Mignot L, Jouenne T, Junter G-A (1989) Appl Microbiol Biotechnol 31: 49
164. Mignot L, Junter GA, Labbe M (1989) Biotechnology Techniques 3: 299
165. Delachapelle S, Renaud M, Vignais PM (1991a) Revue des sciences de l'eau 4: 83 (in French)
166. Delachapelle S, Renaud M, Vignais P (1991b) Revue des sciences de l'eau 4: 101 (in French)
167. Sasikala K, Ramana CV, Rao PR, Kovacs (1993) Anoxygenic phototrophic bacteria: Physiology and advances in hydrogen production technology. In: Adv. Applied Microbiology. Hidleman S, Laskin AI (eds) Academic, San Diego, p 211
168. Khan MMT, Bhatt JP (1991) Int J Hydrogen Energy 16: 83
169. Khan MMT, Adiga MR, Bhatt JP (1992) Int J Hydrogen Energy 17: 93
170. Matsunaga T (1991) Production of useful chemicals from CO₂ by marine microalgae using a biosolar reactor employing light diffusing optical fibres. In: 2nd Int. Marine Biotechnology Conf. Baltimore, MD, USA. 13–16 October 1991, p 61
171. Pohl P, Kohlhase M, Martin M (1988) Photobioreactors for the axenic mass cultivation of microalgae. In: Stadler T, Karamanos Y, Mollion H, Morvan H, Verdus M-C, Christiaen D (eds) Algal Biotechnology. Proc. 4th International Meeting of the Society for Applied Algology. Villeneuve d'Ascq (France), p 209
172. Linton EA, Knowles CJ, Bunch AW, Higton G (1987) The biotechnological potential of microbial hollow-fibre bioreactors. In: Proc. International Conference on Bioreactors and Biotransformations. Gleneagles, Scotland, UK, 9–12 November, 1987, p 299
173. Bunch AW (1988) J Microbiol Methods 8: 103
174. Donoghue C, Brideau M, Newcomer P, Pangrle B, DiBiasio D, Walsh E, Moore S (1992) Use of Magnetic resonance imaging to analyze the performance of hollow-fibre bioreactors. In: Biochemical Engineering VII. Annals of the New York Academy of Sciences, 665, New York, p 285
175. Nyberg SL, Shatford RA, Peshwa MV, White JG, Serra FB, Hu WS (1993) Biotechnol Bioeng 41: 194
176. Feder J (1987) Flat bed hollow fibre reactors for large-scale cultivation of animal cells. In: Mosbach (ed). Methods in Enzymology. Academic Press, Harcourt Brace Jovanovich, Orlando, vol 135, p 393
177. Chiemchaisri C, Yamamoto K, Vigneswaran S (1993) Water Science and Technology 27: 171
178. McKinley KR, Rocheleau RE, Takahashi PK, Bulina EJ, Jensen CM (1992): Hydrogen research and development in Hawaii. In: Veziroglu TN, Derive C, Pottier J (eds) Hydrogen Energy Progress IX. Proc. 9th World Hydrogen Energy Conference, Paris 1992, MCI, Paris, p 641
179. Kentemich T, Haverkamp G, Bothe H (1991) Z Naturforsch 46c: 217
180. Hall DO, Rao KK, Park IH (1991) Immobilized photosynthetic system for the production of fuels and chemicals. In: Environmental Biotechnology. Proc. International Symposium of Biotechnology. Bratislava, 27–29, June 1990, p 259
181. Richmond A (1992) Mass culture of cyanobacteria. In: Mann N, Carr NG (eds) Photosynthetic Prokaryotes. Plenum, New York, p 181
182. Kuwada Y, Ohta Y (1987) J Ferment Technology 65: 597

Synthetic Applications of Enzymatic Reactions in Organic Solvents

A.L. Gutman and M. Shapira
Department of Chemistry, Technion – Israel Institute of Technology,
Haifa 32000, Israel

One of the most important properties of enzymes is their ability to catalyze reactions in a stereoselective manner. This has been used for many years by organic chemists, who have exploited enzymes as catalysts in asymmetric synthesis and resolution for the preparation of optically pure compounds. It is now well established that hydrolytic enzymes can function also in organic solvents and can be used for certain types of transformations which are difficult or impossible to do in water. The present review surveys recent publications of preparatively useful transformations catalyzed by hydrolytic enzymes in nonaqueous media, which enable one to obtain a wide range of homochiral molecules. These transformations exploit the enzymes' enantioselectivity as well as prochiral selectivity. While in the former case the enzymatic reactions amount to a kinetic resolution with the yield of the desired enantiomer not higher than 50%, in the latter case it is often possible to convert a symmetrical prochiral molecule into a homochiral molecule in much higher chemical yields. In addition to the preparative aspects of nonaqueous enzymology, this review also surveys the recent literature related to some fundamental questions of this phenomenon, such as the effect of the nature of the organic solvent and of its water content on reaction rate and stereospecificity of the enzymatic reaction.

Advances in Biochemical Engineering/
Biotechnology, Vol. 52
Managing Editor: A. Fiechter
© Springer-Verlag Berlin Heidelberg 1995

List of Symbols and Abbreviations

Bn	benzyl
CAL	*Candida antarctica* lipase
CCL	*Candida cylindracea* lipase
CVL	*Chromobacterium viscosum* lipase
Da	Dalton
E	enantiomeric ratio
ee	enantiomeric excess
LAPH	liver acetone powder horse
MML	*Mucor miehei* lipase
PCL	*Pseudomonas cepacia* lipase
PEG	poly(ethylene glycol)
PFL	*Pseudomonas fluorescens* lipase
Ph	Phenyl
PPL	procine panacrease lipase
PSL	*Pseudomonas species* lipase
TBDPS	*tert*-butyl-diphenylsilyl
TEA	triethylamine
THF	tetrahydrofurane
TMS	trimethylsilyl
Z	benzyloxycarbonyl

1 Introduction

The importance of optically pure materials is well established nowadays. The understanding that in many cases only one enantiomer is responsible for the biological activity, while the other one may be less active or not active at all, has led to extensive research focusing on various ways of producing homochiral molecules. It has also been shown that the unwanted enantiomer can have undesirable side effects [1, 2], and indeed it is considered in certain countries to be an impurity according to new registration constraints.

During the last decade thousands of papers dealing with the preparation of optically active compounds via enzyme-mediated reactions have demonstrated the usefulness of biotransformations to organic synthesis [3–15]. Of particular interest is the use of hydrolytic enzymes i.e lipases, esterases and proteases, which are commercially available at moderate prices and can be used without additional cofactors. These enzymes have the advantage of accomodating substrates of varying sizes, functional groups and stereochemical complexities.

The realization that enzymes can function in nonaqueous media has eliminated many disadvantages that were associated with their use as catalysts in aqueous environment [16–20]. The main benefits of employing enzymes in organic solvents are summarized below:

The stability of enzymes in nonaqueous media may be greatly enhanced compared to that in water [21, 22]. Water insoluble lipophilic substrates can be transformed more easily to the desired products by working in organic solvents. Compounds that are unstable in water, due to processes like hydrolysis of labile groups or racemization, can be used as substrates in nonaqueous media [23]. The ease of operation is another advantage – the insoluble catalyst can be removed simply by filtration and as a consequence the immobilization of enzymes is often unnecessary. Enzyme properties such as substrate specificity [24, 25] or enantioselectivity [26, 27] can be modified by appropriate choice of solvent. Finally, work in organic solvents has the benefit of shifting the thermodynamic equilibrium of many processes to favor synthesis over hydrolysis [28].

It will be the aim of this review to give a survey of valuable transformations catalyzed by hydrolytic enzymes in nonaqueous media. These transformations exploit the enzymes' prochiral selectivity as well as enantioselectivity and regioselectivity. We will limit our survey to reactions that are conducted in anhydrous organic solvents or in organic solvents that contain up to a few percent of added water. The effect of organic solvent on the reaction rate and the stereospecificity will also be discussed.

2 Exploiting the Enzymes' Prochiral and *meso* Selectivity

Enzymes are widely used as catalysts in organic synthesis. The majority of applications have dealt with the asymmetrization of prochiral and *meso* compounds or with kinetic resolution of racemic substrates [20]. All the kinetic resolution experiments exploit the enzymes' enantioselectivity, i.e., their ability to discriminate between enantiomers of a racemic mixture. The theoretical yield of the wanted chiral product from such reactions is 50%, although in practice it will be considerably lower, since the unfavored enantiomer usually also reacts, albeit at a slower rate. This problem of losing 50% of material may be avoided when it is possible to exploit the enzymes' prochiral or *meso* stereospecificity, i.e. their ability to discriminate between enantiotopic groups or faces of a prochiral (or *meso*) molecule. Several lipases and proteases were shown to exhibit prochiral selectivity under aqueous conditions, for example when hydrolyzing symmetrical C-3-substituted glutarate diesters 1 to the corresponding monoesters 2 (Scheme 1 [29]). It was rewarding to realize that enzymes can also exhibit prochiral selectivity in organic solvents, for example in the acylation reaction of symmetrical C-2 substituted 1,3-propanediols (Scheme 2 [30]).

Like any other catalytic transformation, these reactions are – at least in principle – equilibrium reactions. There are several ways of forcing the reaction in the desired direction. For example, in an esterification reaction the liberated water can be absorbed on molecular sieves [31]. In a transesterification reaction, if a good leaving group is present in the acyl donor, the equilibrium will be shifted toward the products; therefore, the use of activated esters like trifluoro or trichloroethyl esters is very common [32–34]. Oxime esters [35] and anhydrides [36] are also used for the same purpose. The reactions can be made irreversible by conducting the transesterifications with enol carboxylates

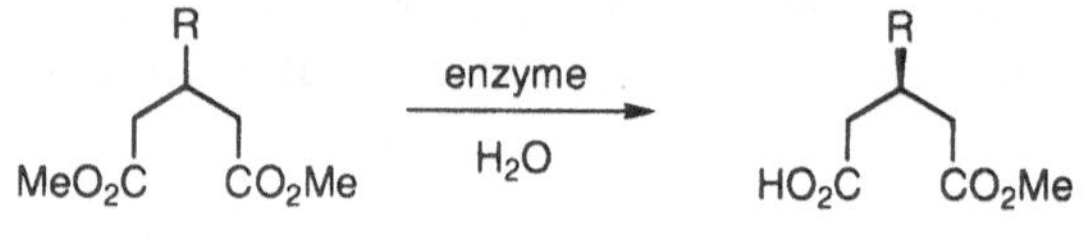

Scheme 1

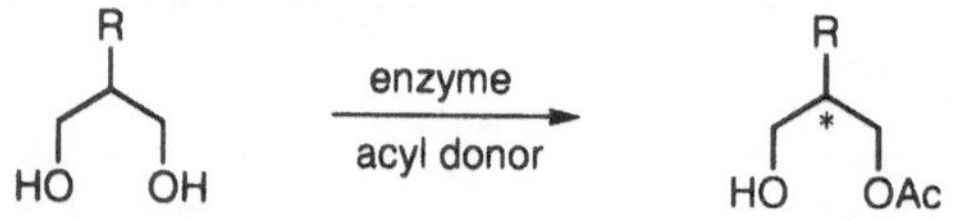

Scheme 2

[37, 38]; the released vinyl alcohol irreversibly tautomerizes to an aldehyde or ketone, and cannot compete for the reverse reaction.

2.1 Asymmetrization of Prochiral Diols

Chiral glycerol derivatives and 1,3-propanediol derivatives have been among the most widely used synthons, and were traditionally prepared by degradation and/or transformation of chiral natural products [39]. These chiral compounds are useful building blocks for the preparation of enantiomerically pure and biologically active molecules such as phospholipids [40], PAF (platelet-activating factor) and its antagonists [41, 42], phospholipase A2 inhibitors [43], renin inhibitors [44] and sphingoglycolipids [45]. In order to prepare these compounds in optically active form it is possible to exploit the enzymes' prochiral selectivity, whereby the enzyme reacts selectively with one of the two prochiral groups on the achiral 2-substituted 1,3-propanediol substrate 3 (Scheme 2).

Table 1 collects some of the studies that were aimed at preparing the above mentioned chiral compounds 4. It can be seen that both the substituent in position 2 and the enzyme used affect the stereochemical outcome of the reaction to give R or S monoester. Most of the products were obtained in excellent yield and optical purity. In comparison with earlier synthetic methods, this route provides efficient access to chiral glycerol compounds without the need for any protecting groups.

The utility of the products 4(a–j) as chiral building blocks is demonstrated, for example, by the synthesis of (S)-propranolol hydrochloride, one of the

Table 1. Asymmetrization of prochiral 2-substituted 1,3-propanediols

Sub-strate	R	Enzyme	Solvent	Config-uration	Yield[1] (%)	ee (%)	Ref.
3a	–Ph	PPL[2]	methyl acetate	R	98	92	[30]
3b	–Bn	PSL	vinyl acetate	R	> 95	> 94	[46, 47]
3c	–CH$_2$-1-naphthyl	PSL	vinyl acetate	R	95	90	[46]
3d	–2,4-dichloro-benzene	PPL[3]	ethyl acetate	R	95	99	[48]
3e	p-BrC$_6$H$_4$CH$_2$CH$_2$–	PPL[2]	methyl acetate	R	90	85	[49]
3f	–OBn	PSL	CHCl$_3$/isopropenyl acetate	S	53	96	[38, 50]
3f	–OBn	PSL	vinyl acetate	S	92	94	[51]
3g	–Me	PFL	CHCl$_3$/vinyl acetate	S	40	> 98	[52]
3h	–CH$_2$CH$_2$CH=CH$_2$	PPL[2]	methyl acetate	R	70	90	[30]
3i	–OEt	PSL	vinyl acetate	S	90	90	[51]
3j	–NHZ	PPL	THF/Vinyl butyrate	R	77	97	[38]

[1] Some authors do not distinguish between conversion and chemical yield, therefore in some cases these numbers may be misleading.
[2] PPL absorbed on Hyflo Super Cel.
[3] PPL precipitated from phosphate buffer/acetone solution and absorbed on Celite

aryloxypropylamine-type β-blockers, from (S)-(+)-2-O-benzylglycerol-1-acetate 4f in several chemical steps [51]. Another example is the use of the chiral products 4b and 4c as key intermediates in the four steps synthesis of potent BW-175 and BW-262 renin inhibitors [47].

It should be mentioned, however, that instead of acylating a prochiral diol, the reaction can be carried out in reverse, i.e. the corresponding diacetate is employed for the asymmetric hydrolysis reaction. As enzymes generally retain their stereoselectivity in organic solvents [21, 32], the two processes are complementary, allowing the preparation of either enantiomer by selecting the reaction conditions and the appropriate substrates (diols or diacetates).

Both enantiomers of 5 (Scheme 3) can be synthesized by monotransesterification of the prochiral 2-sila 1,3-propanediol with methyl isobutyrate or acetoxime isobutyrate using either CCL or CVL (70–76% ee, up to 80% yield) [53]. The products are synthetically useful silyl-chiral optically active compounds. PSL- catalyzed asymmetric esterification of 2,3,6,6-tetrakis(hydroxymethyl) spiro[3.3]heptane with vinyl acetate in both isopropyl ether and pyridine (1 : 1, v/v) gave 2,6-bis(acetoxymethyl)-2,6-bis(hydroxymethyl)spiro[3.3]heptane 6 (Scheme 3) having axial chirality with moderate optical purity (59% ee, 60% yield) [54].

2.2 Asymmetrization of Prochiral Diesters and Hydroxyesters

4-Aryl-1,4-dihydropyridine compounds 7,8 have been widely used for the treatment of cardiovascular diseases since 1975. There are more than 50 derivatives under pharmacological and clinical development, some of which have already been employed therapeutically. Different substituents in these compounds lead to chiral derivatives possessing an asymmetric carbon at the 4-position, and the two enantiomers have been reported to show different biological activities. Although most of these compounds were developed and employed as racemic mixtures, some chiral derivatives have recently come into use [55]. The biocatalytic approach to these compounds includes the enantiotopically-selective hydrolysis of the prochiral diesters 7 to the corresponding chiral monoesters 8 catalyzed by enzyme in organic solvent saturated with water (Scheme 4). The best results are summarized in Table 2.

The discrimination between the two enantiotopic carbonyl groups of 3-substituted glutaric anhydrides 9 is reported to afford the corresponding chiral monoester (Scheme 5 [57, 58]). Several glutaric anhydrides with different substituents at the position 3 were submitted to the lipase-catalyzed reaction in

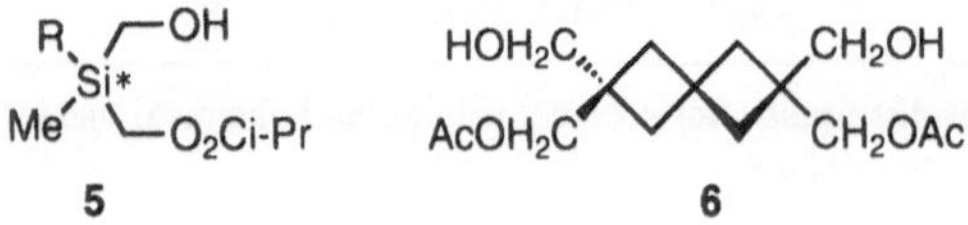

Scheme 3

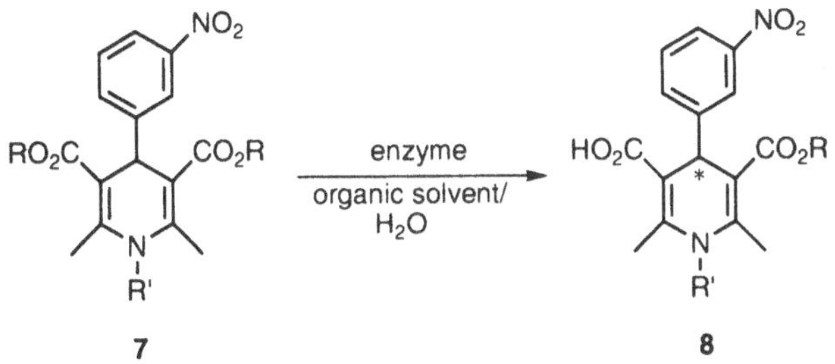

Scheme 4

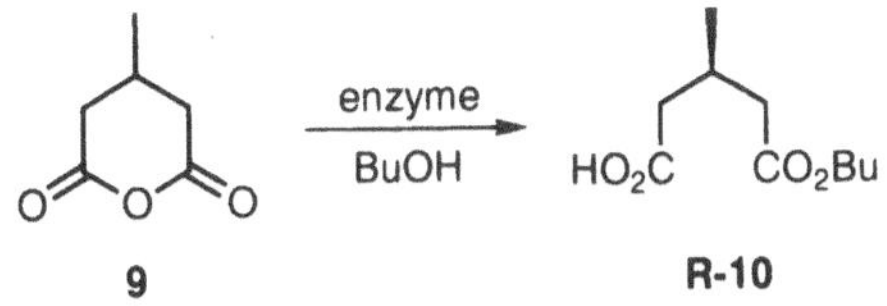

Scheme 5

Table 2. Lipase catalyzed hydrolysis of diesters **7** in organic solvents saturated with water

Substrate	R	R'	Enzyme	Solvent	Conf.	Y (%)	ee (%)	Ref.
7a	$-CH_2O_2CCH_3$	Bn	PSL	$BuOH:H_2O, 10:1$	S	81	97	[56]
7b	$-CH_2O_2Ct$-Bu	H	PSL	$i\text{-}Pr_2O/H_2O^2$	S	87	> 99	[55]
7c	$-CH_2O_2CC_2H_5$	H	PCL	$i\text{-}Pr_2O/H_2O^2$	R	86	99	[55]
7d	$-CH_2O_2CCH_3$	H	PCL	$i\text{-}Pr_2O/H_2O^2$	R	87	> 99	[55]

[1] Isolated as the methyl ester after reaction with diazomethane.
[2] Saturated with H_2O

organic solvents with 1-butanol. Best results for the nucleophilic ring-opening are obtained for the prochiral 3-methyl glutaric anhydride with PFL in diisopropyl ether and with butanol as the nucleophile. The (R)-butyl ester **10** is obtained with 74% yield and 91% ee [57]. In another example, chiral 5,5-disubstituted N-acyloxymethylbarbiturates **12** have been obtained in 40–99% optical yields and up to 62% chemical yields by lipase-catalyzed hydrolyses of prochiral 5,5-disubstituted bisacyloxymethylbarbiturates **11** in H_2O-saturated diisopropyl ether (Scheme 6 [59]). These chiral barbiturates were readily converted into the corresponding chiral N-methylbarbiturates which have various pharmacological activities.

Lactonic functionality is present in a large variety of natural products and biologically active compounds. Lactone derivatives are very common flavor components used in the perfume and food industry. They have also been reported to be sex attractant pheromones of different insects and to be plant-growth regulators. Most of them are chiral, and their physiological activity often

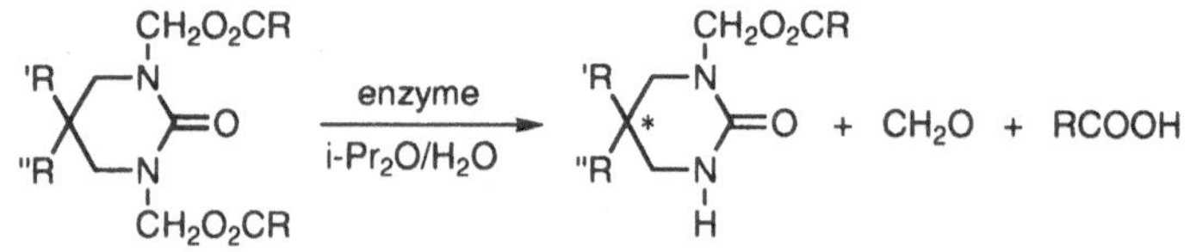

Scheme 6

depends on the absolute configuration. Owing to the importance of this class of compounds, many optically active lactones have been the targets of an increasing number of synthetic efforts, in recent years. Intramolecular transesterification of the symmetrical γ-hydroxypimelate diester 13 catalyzed by two different enzymes PPL or PFL in hexane leads to the chiral lactone (S) or (R)-14 in an enantioconvergent manner (Scheme 7, Table 3 [60–62]). The stereospecificity of the two enzymes is markedly different. PPL gives the (S) enantiomer with high ee, whereas PFL produces the (R) configuration in lower optical yield. The lactone 14d was further reduced to (S)-(−)-γ-lactonic acid and crystallized in quantitative yield with > 95% ee.

It is noteworthy that this approach can work only in organic solvents and is not possible in aqueous solutions, where enzymatic hydrolysis of the lactone 14 would produce a symmetrical molecule resulting in the loss of chirality. Another example where work in organic solvents is an absolute necessity is in the preparation of chiral monosubstituted malonates 16. The failure to prepare these compounds in optically active form has been due to the inapplicability of the conventional approach, i.e. enzymatic hydrolysis of the symmetrical monosubstituted diesters to the corresponding half esters. Although successfully used for the synthesis of chiral disubstituted malonates [63–65], this approach was bound to fail for the monosubstituted malonates , because under aqueous

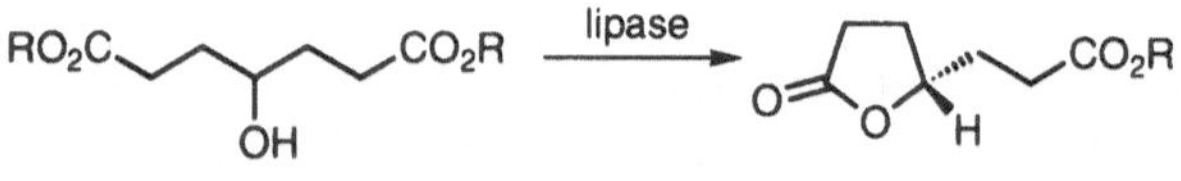

Scheme 7

Table 3. Enantioconvergent lactonization of γ-hydroxypimelates with lipases in hexane [60]

Prochiral substrate	R	Enzyme	Conv. (%)	Product	ee (%)
13a	Me	PPL	100	S-14a	> 95
13b	Et	PPL	100	S-14b	> 98
13b	Et	PFL	100	R-14b	32
13c	i-Pr	PPL	33	S-14c	46
13c	i-Pr	PFL	36	R-14c	10
13d	Bn	PPL	100	S-14d	> 95

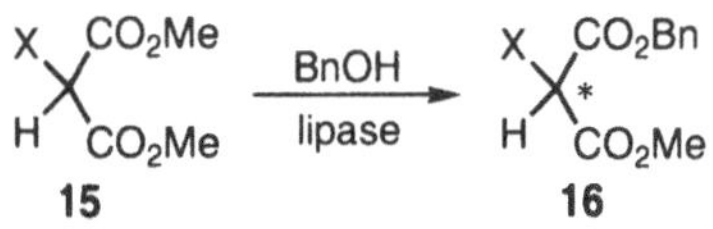

Scheme 8

conditions, the activated malonic hydrogen undergoes fast exchange leading to racemization. The prochiral stereospecificity of enzymes in organic solvents is used for the formation with a high degree of optical purity of chiral mixed diesters of 2-substituted malonic acid **16** with different alkyl and aryl substituents in the 2-position [66, 67]. The reaction involves enzymatic transesterification of symmetrical monosubstituted dimethyl malonates **15** with benzyl alcohol in hexane (Scheme 8). The synthetic utility of this method is further demonstrated by converting the chirally unstable mixed methyl benzyl diesters to the corresponding half esters, which are in turn selectively reduced to chirally stable and synthetically useful hydroxyesters [67].

The same strategy has been used for the enantioconvergent polymerization of prochiral dimethyl β-hydroxyglutarate to give chiral oligomers with modest ee values (30–37%) [68]. An important feature of such chiral polymers is that they possess functional groups, which may be used for cross-linking and as anchors for attaching achiral reagents.

2.3 Asymmetrization of meso Diols

The group of *meso-cis*-diols serves as another example where the maximum theoretical yield of the desired chiral product can reach 100%. Most of the work has been focused on the transesterification of cyclic *meso-cis*-diols, and in Table 4 some of these works are outlined.

Enantiomerically pure **17** or its enantiomer are chiral starting materials for the synthesis of prostaglandins and other cyclopentanoid natural products. Optically pure **17** was prepared from the corresponding *meso*-diol by lipase-catalyzed transesterification using vinyl acetate as acyl donor and THF/TEA as organic solvents in 65% yield [69, 81]. The enantiomer of **17** was synthesized starting from the corresponding *meso*-diacetate (in a mixture with the racemic *trans*-diacetate) by the alcoholysis reaction using PPL in ethanol and hexane (53%, > 98% ee) [82]. Chiral **18** can serve as a valuable precursor to optically-active cyclohex-1-en-3-ol, a chiral building block not readily accessible, and also as a precursor to 4-acetoxycyclohex-2-en-1-one which serves as an intermediate in the synthesis of Mevinolin and Compactin as well as the immunosuppressant FK506 [70]. Very good results were also obtained for the 7-membered ring **19** (81% yield, > 99% optical purity) [71], and the 6-substituted 7-membered ring **20**, which can be further utilized in the synthesis of skipped polyols and sugar derivatives [72].

The preparation of optically active **21** and related compounds with 3, 4 or 6-membered rings was successfully accomplished using PSL. Whereas slow

Table 4. Asymmetric transesterification of *meso-cis*-Diols

	Product	Enzyme	Solvent	Y (%)	ee (%)	Ref.
17	AcO⟋⟍OH (cyclopentene)	Pancreatin	Vinyl acetate THF/TEA	65	> 99	[69, 81]
18	HO···⟨⟩···OAc (cyclohexene)	PSL	isopropenyl acetate	51	95	[70]
19	AcO⟋⟍OH (cycloheptene)	CAL[1]	isopropenyl acetate	81	> 99	[71]
20	OTBS AcO⟋⟍OH (cycloheptene)	PSL	isopropenyl acetate	95	> 95	[72]
21	OH OAc (cyclopentane)	PSL	vinyl acetate	85	> 95	[73]
22	HO⟋⟍OAc (acetonide)	PFL	vinyl acetate/ Et₂O	51	> 98	[74, 75]
23	AcO⟋O⟍ HO⟋O	PSL	vinyl acetate	80	95	[76]
24	OAc OH (bicyclic)	Pancreatin	vinyl acetate THF/TEA	94	> 99	[69]
25	O OAc O OH	PSL	vinyl acetate	81	> 99	[77]
26	AcO··⟨⟩··OH H CO₂Et (decalin)	PFL	vinyl acetate	96	> 99	[78]
27	AcO H HO·· H	PSL	vinyl acetate/ MeCN	87	> 99	[79]
28	O OAc OH	PPL	vinyl acetate	92	> 99	[80]

[1] Supplied as an acrylic supported biocatalyst

reaction rates were observed using ethyl acetate as an acyl donor, the reactions were about ten times faster when vinyl acetate was employed [73]. Compound **22** was transformed into one of the mevinic acid analogs in a four steps synthesis as well as to the C-10 fragment of Nystatin A$_1$ macrolide antibiotic [74, 75]. The chiral product **23** can be used as a key intermediate in the synthesis of biologically active molecules such as arachidonic acid metabolites and nucleosides [76], similarly, **25** is converted into the natural carbocyclic nucleoside aristeromycin [77].

By conducting a transesterification reaction with vinyl acetate in THF/TEA and using pancreatin, the monoacetate **24** can be obtained from the corresponding *meso*-diol in high chemical and optical yields (94%, > 99% ee) [69]. The same *meso*-diol can be used as a substrate in a lipase-catalyzed "doubly enantioselective" transesterification with racemic trifluoroethyl 2-chloropropanoate to give four diastereomers [83]. Analysis of the results shows that the selectivity of the first step of this transformation - the enantioselection between the enantiomers of the racemic ester - is very low. The second step in this transformation, the acyl transfer from the acyl enzyme to the enantiotopic groups of the *meso*-diol, is highly selective, giving two diastereomers almost exclusively with very high ee values (93–100%). The described process combines asymmetrization of a *meso*-diol with the resolution of a racemic ester to give a chiral molecule with three asymmetric centers in one step.

The monoacetate **26** can serve as a chiral building block for the enantioconvergent synthesis of sesquiterpenes involving a drimane ring skeleton like (−)-polygodial, (−)-warburganal and (−)-drimenin [78]. Other important intermediates are **27** for the synthesis of conduritol C [79] and **28** which is a synthetically useful, optically active Diels-Alder type derivative [80]. Other examples of enzymatic asymmetrization of *meso*-diols in organic solvents have been reported [84–88].

The use of biotransformations for the preparation of homochiral organometallic transition metal complexes has also been observed [89, 90]. For instance, acetylation of 1,2-bis(hydroxymethyl)ferrocene **29** with vinyl acetate in benzene affords (1S)-(−)-2-acetoxymethyl-1-0hydroxymethylferrocene **30** or its antipode, as optically pure compounds in 80 or 57% yield respectively, when immobilized PSL or CVL are used as catalysts (Scheme 9 [89]).

The *meso* diesters or anhydrides can also serve as substrates in enzymatic asymmetrization. The chiral *cis*-disubstituted aziridine **31** (Scheme 10) can be prepared by means of enzymatic transesterification of the corresponding meso-diacetate [91]. The reaction involves treating the *meso*-diacetate with BuOH in

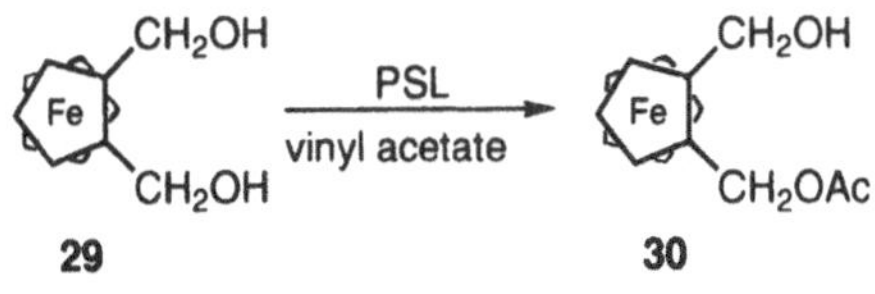

Scheme 9

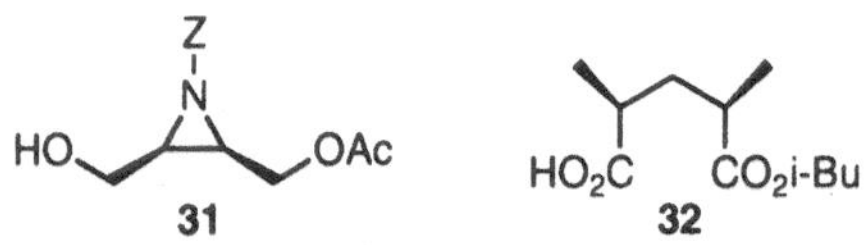

Scheme 10

i-Pr$_2$O in the presence of enzyme. PSL gives the best results affording **31** in 98%
ee and 68% yield. This type of transesterification is sometimes referred to as
alcoholysis or the deacetylation reaction. The chiral aziridine moiety can serve
as a key intermediate in the synthesis of different antibiotics like mytomycin C
which has already been used for the cancer therapy [91]. The chiral monoester
32 (Scheme 10) is another important building block for various applications
(polyether antibiotics, (−)-α-multistriatin pheromone and others). This monoester is available by asymmetric alcoholysis of *meso*-2,4-dimethyl-
pentanedioic anhydride [92]. The best results are obtained using lipase from
Candida cylindracea (CCL) with i-BuOH in cyclohexane (72%, 90% ee).

3 Exploiting the Enzymes' Enantioselectivity

The resolution of racemates can be achieved by making use of the ability of
enzymes to discriminate between enantiomers of a racemic mixture. A requirement for successful resolution is sufficiently different rate constants for the
transformations of the enantiomers. As was mentioned before, the theoretical
yield of the desired chiral product from such reactions is 50%, although usually
the reaction is stopped at low (≤ 45%) or high (≥ 55%) conversion in order to
obtain the product or substrate respectively, with high ee. The enantioselectivity
of the biocatalytic reactions is normally expressed as the enantiomeric ratio E, a
biochemical constant that is independent of substrate concentration and the
extent of conversion [16, 33]. The E value is a measure of the enzyme's
discrimination between two competing enantiomers, and is the ratio of the rate
constants for the fast and slow enantiomers.

3.1 Resolution of Racemic Amines

Although the most common enzyme-catalyzed reactions are stereoselective
esterifications and transesterifications, it has also been reported that some
lipases and proteases are effective in catalyzing the reaction between carboxylic
esters and amines in anhydrous organic solvents [32]. These reactions have been
used in the resolution of racemic amines as well as in the synthesis of peptide
bonds (that will not be discussed in this review).

Chiral amines are of great synthetic utility for the pharmaceutical and fine-chemical industries [93]. They are very common as intermediates and end products in the synthesis of many drugs and as chiral resolving agents for the preparation of chiral carboxylic acids. Since naturally occurring compounds containing an amide bond frequently show biological activity, the synthesis of optically active amides is also an area of growing interest. Enzyme-catalyzed aminolysis in organic solvents may provide a useful and general alternative for the preparation of optically active amines and amides. Several of the resolved amines are collected in Scheme 11.

Mainly two enzymes are used, the protease subtilisin that predominantly gives the S-amide and the CAL that give the R-amide in preference. Compound 33a is resolved successfully with different methyl 3-arylpropanoates with CAL in diisopropylether, obtaining R-34 in 42% yield and > 95% ee [94]. S-34 can be prepared using subtilisin with trifluoroethylbutyrate in anhydrous 3-methyl-3-pentanol (33%, 85% ee [27]). The same compound 33a is resolved with vinyl carbonates to give the optically active carbamate derivatives. Best results are obtained with n-octyl vinyl carbonate and CAL in hexane or diisopropylether to give R-34 with 98% ee [95]. Carbamate derivatives are products of considerable interest in some areas of medicinal chemistry. An alkaloid, physostigmine, which has a carbamate moiety in its structure, is currently undergoing clinical trials for the treatment of Alzheimer's Disease. The use of vinyl carbonates is also applied to compounds 33c,d (40–62% ee) [95], but these compounds are resolved preferably using methyl acrylic esters as acyl donors in THF with CAL (20–27% , 95% ee) [96]. Such α,β-unsaturated amides can be used for the synthesis of chiral polymers as is demonstrated, for example, in the case of 33b. First, substilisin is used for the stereoselective acylation with trifluoroethyl methacrylate in 3-methyl-3-pentanol to give S-34 (18%, > 98% ee). Second, the resultant monomer is chemically polymerized to form an optically active polymer with molecular weight of 10^5 Da and 96% ee [97].

Subtilisin immobilized on glass beads was used for the production of the pharmaceutically important intermediate (R)-1-aminoindan 36 (Scheme 12) with trifluoroethyl butyrate in 3-methyl-3-pentanol. The reaction was carried out in a continuous-flow column bioreactor [98]. At steady-state operating conditions, a column constructed with 3.7 g of enzyme on 370 g glass beads

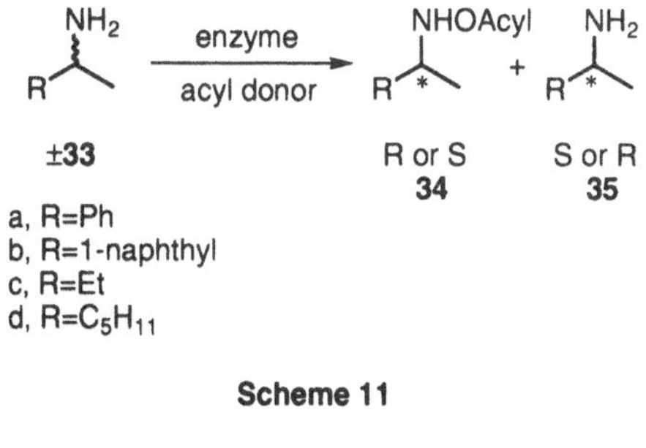

±33

a, R=Ph
b, R=1-naphthyl
c, R=Et
d, R=C$_5$H$_{11}$

R or S
34

S or R
35

Scheme 11

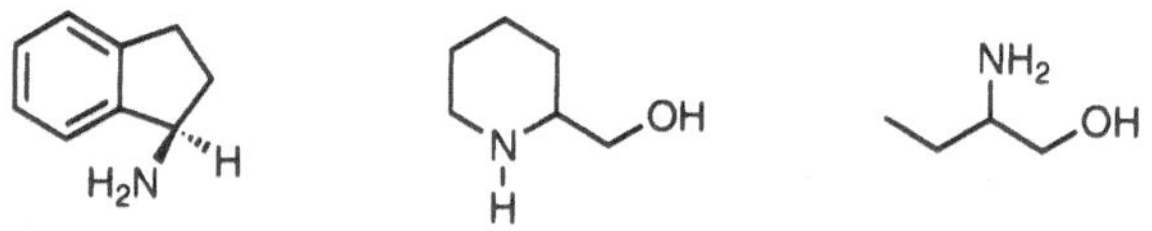

Scheme 12

operated continuously for 95 h to elute 330 g of racemic aminoindan and to give in 40% yield (*R*)-1-aminoindan of optical purity higher than 98%. The acylation of the racemic amino alcohol **37** with ethyl acetate and PPL at 0–5°C to give the *S*-hydroxy amide (9%, 92% ee) is a rare example of resolving a secondary amine [99]. Under the conditions mentioned above, no acylation occurred on the hydroxy group.

The amino alcohols like compound **38** represent important structural features of natural products such as adrenaline, β-adrenergic receptor blockers and local anaesthetics. Racemic amino alcohol **38** can be resolved through the amine moiety by PPL-catalyzed acylation with ethyl acetate to produce the *S*-amido alcohol with 95% ee and equal amount of the *R*-amido ester with 95% ee as well [100]. The same compound can be resolved through the alcohol moiety by using the benzyl carbamate derivative as substrate for the transesterification reaction [100], or by using the amido ester as substrate for the alcoholysis reaction with butanol [101].

3.2. Resolution of Racemic Acids

Asymmetric esterifications catalyzed by lipases in organic solvents can be used for the resolution of racemic acids [102]. Most of the work has focused on CCL-catalyzed esterification of 2-substituted carboxylic acids, as illustrated in Scheme 13 and Table 5.

The process is successfully employed for stereoselective esterifications of several 2-halo-carboxylic acids, like compounds **39a,b**. High optical purities are obtained in most cases, and the *R*-esters are formed. Both **39a** and the corresponding 2-chloro substrate are used for the synthesis of a number of derivatives of 2-phenoxypropanoic acid, having wide applications as potent herbicides. The resolution of these two carboxylic acids is currently being

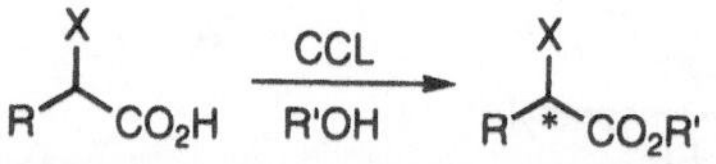

Scheme 13

Table 5. Resolution of racemic acids by asymmetric esterification

product	R	X	R' (solvent)	conv. (%)	ee (%)	E	Ref.
40a	CH₃	Br	n-butyl (hexane)	45	96	–	[103]
40b	C₄H₉	Br	n-butyl (hexane)	30	99	–	[103]
40c	C₃H₇	Me	n-octyl (heptane)	46	93	70	[104]
40d	C₄H₉	Me	octadecyl (heptane)	50	84	30	[104]
40e	C₆H₁₃	Me	hexadecyl (cyclohexane)	42	97	> 100	[105]
40f	p-Cl-PhO	Me	cyclohexyl(i-octane¹)	45	> 95	80	[106]

¹ Saturated with phosphate buffer, pH = 7

carried out on a 100-kg scale and is to be scaled up further by Chemie Linz Co. (Austria) under a license from MIT [18]. Optically pure 2-methylalkanoic acids and their derivatives can serve as valuable synthetic intermediates in the preparation of, among other compounds, a number of insect pheromones. These compounds can be resolved using CCL-catalyzed esterification with various alcohols in organic solvents. For instance, compounds **39c–e** (Table 5) were esterified to the corresponding **S-40c–e** with ee values of 84–97%. In the resolution of 2-methyloctanoic acid **39e**, the E values increased continually with increasing chain length of the alcohol used [105].

The resolution of racemic 2-(4-chlorophenoxy)-propanoic acid **39f** is achieved with cyclohexanol in isooctane saturated with phosphate buffer at pH = 7 to produce **R-40f** (ee > 95% at 45% conv.) [106]. Compound **39f** can also be resolved in a continuously operated biochemical system. The long-term continuous resolution of **39f** is carried out by stereoselective esterification with celite-absorbed lipase from CCL using n-tetradecanol as the second substrate in water saturated carbon tetrachloride-isooctane (8:2, v/v) [107]. Under optimized conditions the R-enantiomer of the acid is continuously esterified with high stereoselectivity (ee = 95%) in a packed-bed column reactor for 34 days.

Another approach to the same resolution is the 'doubly enantioselective' esterifications using racemic acids and racemic alcohols as substrates. Ester formation from the racemic acid **39f** and the racemic alcohol **41** can obviously lead to four stereoisomers of **42** (Scheme 14). The CCL-catalyzed esterification gives predominantly the **(R, R)-42** isomer (79%) at 32% conversion [108]. Lithium aluminium hydride reduction of the product gives the alcohol **41** and

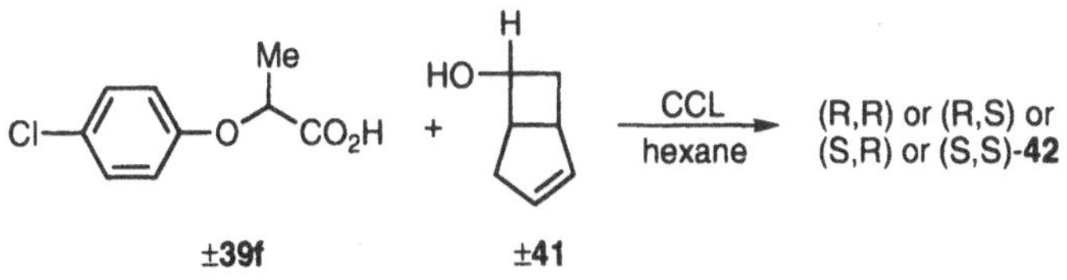

Scheme 14

the alcohol that corresponds to the acid **39f** with 70% and 83% ee respectively. Higher optical purities are obtained using the corresponding acetate instead of the alcohol **41**, thus, the **(R, R)-42** is obtained in 91.5% (13% conversion) and the ee's of the corresponding alcohols, after reduction, are 93% and 88% respectively [108].

3.3 Resolution of Racemic Alcohols

The majority of applications of enzyme-catalyzed resolutions of racemic substrates relates to the resolution of racemic alcohols. With the exception of certain sterically hindered alcohols, hydrolytic enzymes catalyze the asymmetric acylation of a wide range of acyclic and cyclic substrates, including various types of primary and secondary hydroxy compounds. In order to achieve highly enantioselective acylation, it is necessary to select a suitable biocatalyst and to optimize the reaction conditions [16].

Using different lipase preparations with various acyl donors in organic solvents the racemic primary alcohols **43(a–e)** are efficiently resolved into optically active esters **(S)-44(a–e)** and the corresponding unreacted alcohols **(R)-43(a–e)** (Scheme 15, Table 6).

As has already been mentioned, the reaction is stopped at low (≤ 45%) or high (≥ 55%) conversion in order to obtain either the product, in the former case, or the substrate, in the latter, with high ee. 2-Benzylpropanol **43a** or the corresponding acid are potentially useful compounds for incorporating chiral recognition in adenosine receptor agonists and antagonists. The attempts to synthesize this unit via 2-benzylpropionaldehyde, obtained by a Sharpless

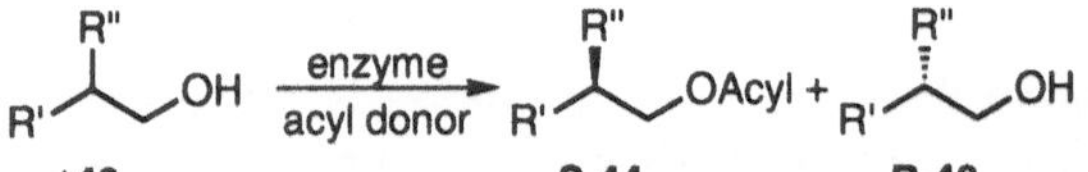

Scheme 15

Table 6. Resolution of primary alcohols

subs-trate	R'	R''	(S)-44			(R)-43			Ref.
			conv (%)	y (%)	ee (%)	conv (%)	y (%)	ee (%)	
43a	Me	–Bn	–	45	74	–	43	97	[109]
43b	Me	–CH₂CH₂SPh	–	–	–	60	45	98	[110]
43c	Me	–CH₂OTBDPS	40	39	> 98	60	38	> 98	[52]
43d	Me	–CH₂OBn	60	–	90	40	–	90	[111]
43e	Me	–BnOBn	45	–	98	55	–	95	[48]

epoxidation procedure, resulted in modest stereoselectivity [112]. Using PSL-mediated transesterification with vinyl acetate in t-BuOMe, **R-43a** is obtained with high optical purity (97% ee, 43% [109]). Transesterification with vinyl acetate using PFL in CHCl$_3$, is applied to racemic **43b**, another important building block, to provide almost optically pure R-alcohol at 60% conversion [110].

Optically pure derivatives of 2-methyl-1,3-propanediol are important synthons for the preparation of many natural products. A biocatalytic approach to the synthesis of these chiral building blocks by the enzymatic transesterification of the prochiral 1,3-diols was discussed in a previous chapter. An alternative is to use racemic 2-methyl-1,3-propanediol derivatives as substrates for enzymatic resolution. This route is utilized in the PFL-catalyzed resolution of racemic **43c,d** with vinyl acetate in CHCl$_3$ to obtain either the ester or the remaining alcohol with high ee (90–98% [52, 111]). Racemic **43e** is resolved with PSL in ethyl acetate and serves as starting material in the synthesis of a morpholine fungicide [48].

An interesting example is the use of 2-phenyloxazolin-5-one **46** as the acylating agent in the resolution of racemic amido alcohol **45** (Scheme 16 [113]). The resolution is carried out with MML in diisopropylether in the presence of 0.4 equivalents of the azlactone. The reaction stops after the complete consumption of **46**, thus avoiding the need to closely monitor the conversion. The solid product **(R)-47** (93% ee), being insoluble in diisopropylether, drops out of the reaction mixture when formed, and hence no chromatography is needed for its separation from the unreacted enantiomer. Other approaches for the resolution of 2-amino-1-butanol derivatives were discussed in Sect. 3.1.

A good alternative to the preparation of optically active epoxyalcohols by the Sharpless asymmetric epoxidation of allylic alcohols is the direct enzyme-catalyzed resolution of the epoxyalcohols. This approach was used for the resolution of racemic **48** (Scheme 17) with PSL and trifluoroethyl butyrate in

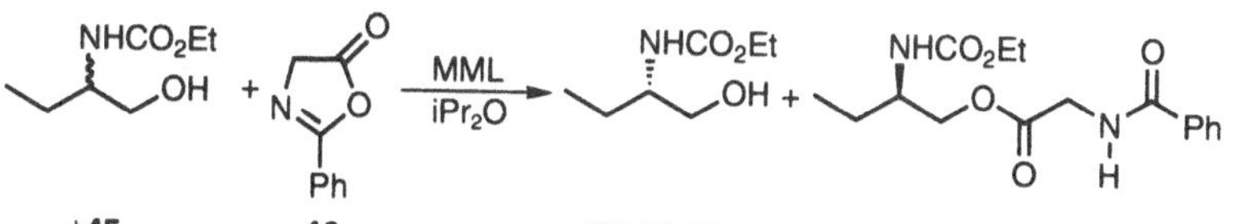

Scheme 16

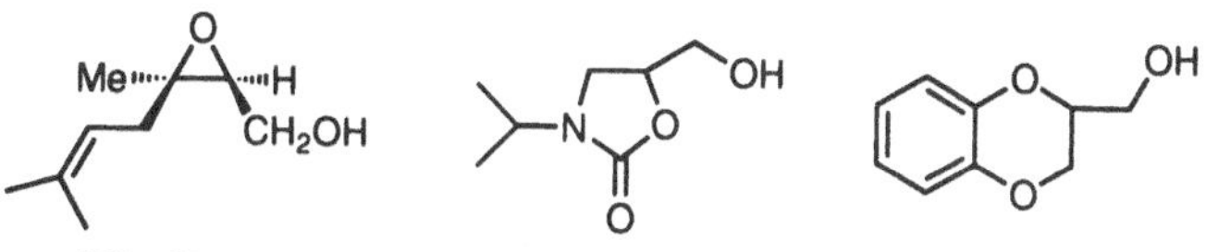

Scheme 17

THF, affording **(2R,3S)-48** with 42% yield and ee > 95% at 55% conversion [114]. Other resolutions of epoxyalcohols have also been reported [115, 116]. High optical purities are obtained in the case of racemic **49** (Scheme 17). The R enantiomer of **49** is preferentially transesterified by PSL (absorbed on celite) with acetic or propanoic anhydride as the acyl donors in benzene. When the reaction is stopped at 50% conversion both the ester and the unreacted alcohol are obtained with ee > 95% (40–45% [36]). Similar compounds, N-hydroxymethyl-γ-butyrolactams, are successfully resolved with PSL and vinyl acetate in t-BuOMe [117].

The 1,4-benzodioxane ring system occupies an important place among cardiovascular agents. A few derivatives of this class of compounds possess antihypertensive properties by virtue of their α-adrenergic antagonism, other derivatives have been found to exhibit strong affinities for other CNS receptors (central nervous system) in particular the serotonergic and dopaminergic sites [118]. The biological activity of these compounds is considerably influenced by the chirality of the 1,4-benzodioxane unit, the S-enantiomer having higher binding activities. 2-Hydroxymethyl-1,4-enzodioxane **50** (Scheme 17) is resolved kinetically in a PFL-catalyzed transesterification with vinyl acetate in dioxane to afford at 62%conversion the remaining **(S)-50** with ee > 99% [119]. Other monocyclic and bicyclic compounds having a primary hydroxyl group have been resolved successfully [118, 120, 121].

Numerous open-chain secondary alcohols have been resolved into their enantiomers using an enzyme-catalyzed acyl-transfer reaction, some of them are listed in Table 7 (Scheme 18).

Table 7. Resolution of open-chain secondary alcohols

Sub-.	R'	R''	ester **52** conv (%)	ester **52** ee (%)	alcohol **51** conv (%)	alcohol **51** ee (%)	E	Ref.
51a	Me	alkyl (C$_2$–C$_{14}$)	17–50	93–100	47–58	90–100	41–100	[38, 103, 122, 123]
51b	Me	–(CH$_2$)$_2$CH=CMe$_2$	27–30	88–98	52–65	97–100	100[1]	[38, 125, 126]
51c	Ph	alkyl (C$_1$–C$_5$)	13–50	91–100	47–56	87–100	25–180	[31, 36, 127–129]
51d	Bn	Me	11–49	90–100	49–56	91–100	24–300	[31, 127–129]
51e	Me	2-naphthyl	37–43	89–100	52–60	99–100	30–460	[31, 127, 128]
51f	Me	cycloalkyl (C$_5$–C$_6$)	29–43	94–96	29–43	40–71	50–70	[31, 129]
51g	Ar	–CH$_2$X (X = Cl, Br)	49–52	92–97	49–52	80–97	–	[133]
51h	–CH$_2$Cl	–CH$_2$OAr	29–50	92–100	50–56	72–99	45–700	[130, 134–137]

[1] The E value was reproted in one case only

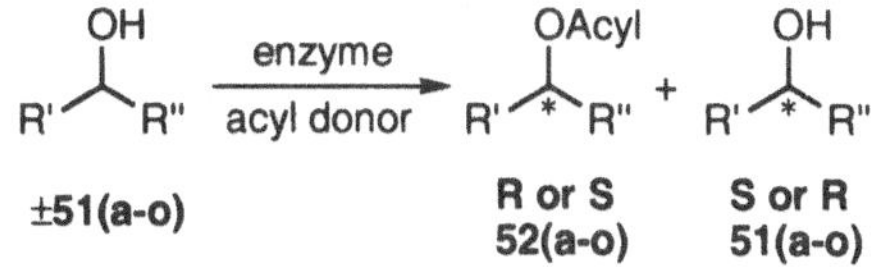

OH enzyme OAcyl OH
R'⁀R" —acyl donor→ R'⁀*⁀R" + R'⁀*⁀R"

±51(a-o)   R or S   S or R
52(a-o)   51(a-o)

Scheme 18

Aliphatic secondary alcohols, like **51a**, constitute an industrially important class of aroma and flavor compounds. Several 2-alkanols **51a** have been resolved using PPL or CAL and various acyl donors (vinyl acetate, trichloroethyl butyrate, trifluoroethyl laurate) to obtain **(R)-52a** or **(S)-51a** with high optical purities. The use of S-ethyl thiooctanoate as acyl donor is also fruitful, leading to both the thioester and the alcohol with ee > 97 at approximately 50% conversion [123]. Reducing the pressure in the reaction mixture is a useful method to achieve high equilibrium conversion in transesterifications. Under reduced pressure the volatile alcohol formed upon transesterification is evaporated, shifting the reaction toward the products. This effect is used to resolve, among other secondary alcohols, 2-octanol with CAL and ethyl octanoate at 15 mm Hg affording the ester and the remaining alcohol with ee ≥ 94 at 52% and 45% conversion, respectively [124].

Sulcatol **51b** is the aggregation pheromone of an ambrosia beetle. The chirality of the molecule is involved in its biological activity: one species responds to a mixture of 65% of isomer **(S)-(+)-51b** and 35% of the **(R)-(−)-51b** and does not respond to either one of these isomers alone. Another species is sensitive only to the S-enantiomer, and its response seems to be inhibited by the R-enantiomer [125]. Very high optical purities are obtained for **(S)-51b** and the corresponding ester **(R)-52b** when racemic sulcatol is resolved with PPL and various activated or enol esters. In a certain study the enantioselectivity of the transesterification was significantly increased by dehydration of the enzyme to a constant weight [126].

Enantiomerically pure alkyl-aryl secondary alcohols (**51c–e** and others) are useful chiral auxiliaries in organic synthesis for both analytical and synthetic applications.These alcohols are conveniently obtained in high optical and chemical yields by enzyme-catalyzed transesterification with different enzymes - PPL, PSL and mucor esterase, with PSL giving slightly better results than the others. Unactivated [31] and activated esters [129] as well as enol esters [128, 130] and anhydrides [36, 127] are used as acyl donors. The use of cyclic anhydride [127] is particularly rewarding since the enzymatic acylation converts one enantiomer of the alcohol into a half acid, which can be separated from the unreacted alcohol by extraction with an aqueous base, thus avoiding the need of separating an ester from the unreacted alcohol by tedious chromatography or distillation. The phenyl and naphthyl groups can be substituted [31, 128, 130] or replaced with other aromatic systems like furan [31] without significant loss in enantioselectivity and yield. The methyl group (in **51c**) can be replaced as well, for instance with a trifluoromethyl [128], piperidine [131] or acylaminomethyl [132] group. Cycloalkyl derivatives **51f** are also reported as substrates for enzymatic resolution to afford **(R)-52f** with high optical yield. Even polycyclic derivatives like adamantylethanol and norbornenylethanol can be submitted to the enzyme, albeit with moderate reaction rates [31].

2-Halo alcohols (**51g, h**) are versatile synthons in organic synthesis. They can be converted into synthetically useful oxiranes upon base treatment, or further elaborated into several β-blockers, lipids and pheromones. For example β-adrenergic blocking agents of the 1-alkylamino-3-aryloxy-2-propanol type (e.g. atenolol, propranolol) can be easily obtained in several chemical steps from the corresponding 1-chloro-3-aryloxy-2-propanols **51h**. It is well established that the desirable therapeutic activity resides mainly in the S-enantiomers of these molecules. It is also known that some of the opposite R-enantiomers display undesirable side effects [134]. Many derivatives of the racemic halohydrines **51g,h** are successfully resolved with PSL and enol esters to obtain **(S)-52g,h** and **(R)-51g,h** [130, 133–137]. The 3-aryl substituent in **51h** can be replaced with 1,3-dioxin-4-one without losing the enantioselectivity of the enzyme (ee > 98% for both the ester and the alcohol) [138]. Many other open chain secondary alcohols have been enzymatically resolved, including alcohols that contain sulfur [139], silicon [140] or tin [141] functionalities.

The Katsuki-Sharpless epoxidations of allylic alcohols [142] constitutes one of the most important developments in asymmetric synthesis during the last decade. The method has also been used for some valuable kinetic resolutions [143], but there are certain restrictions that limit the application of this approach. In particular, it is not applicable to propargylic alcohols and the products must be isolated from significant amounts of catalyst residues [144]. Irreversible, enzyme-mediated acylations in organic solvents can be utilized as an alternative for the resolution of many allylic and propargylic as well as homoallylic and homopropargylic secondary alcohols in a stereoselective manner. Some of these alcohols are listed in Table 8 (Scheme 18).

In all the cases listed in Table 8, the enzyme (usually PSL) preferentially transesterifies the R-enantiomer of the alcohols. Compound **51i** cannot be

Table 8. Resolution of allylic and propargylic secondary alcohols

Substrate	R′	R″	Conv (%)	ee 51 (%)	ee 52 (%)	E	Ref.
51i	Me	$\curlyvee$ Ph	52	> 95	> 95	> 20	[145]
51j	Me	(E) –CH = CH(Ph)	50	> 95	> 95	> 20	[145]
51k	$CH_2 = CH-$	(E) –CH = CH(Ph)	50	> 95	> 95	> 20	[145]
51l	Et	(E) –CH = CHCO$_2$Me	–	> 95	> 95	> 150	[148]
51m	Me	(E) –CH = CHO$_2$SPh	50	> 98	> 95	> 50	[149]
51n	Me	–C ≡ C–Ph	50	> 95	> 95	> 20	[145]
51o	Et	–C ≡ C–Bu	51	> 95	82	> 20	[145]

resolved via asymmetric Sharpless epoxidation because of the deactivating influence of the electron-withdrawing alkene substituent. The phenyl group in **51i** can be replaced with various ketone or ester functionalities [146] or with the cyano group [147], still obtaining high optical yields. The structure of the substrate has an appreciable effect on the enantioselectivity; for instance, when the Z-isomers of **51j,k** were submitted to the enzymatic reaction, instead of the E-isomers, very low optical yields were obtained [129, 145]. It has also been seen that aromatic groups are particularly favorable, whereby the enantioselectivity drops dramatically when the phenyl group in **51j** is replaced with an alkyl group [145]. γ-Hydroxy-α,β-unsaturated esters **51l** or phenyl sulfones **51m** are also very good substrates. The chiral phenyl sulfone **51m** is further utilized as starting material in the iterative construction of polypropanoate segments with four consecutive chiral centers [150]. Acylations of propargylic alcohols with alkyl and aryl substituents (**51n,o**) are all enantioselective, in contrary to the allylic alcohols. Other examples that exist in the literature include the resolution of silyl derivatives [151] and primary allenic alcohols [152].

Another significant class of secondary alcohols comprise the cyclic and bicyclic hydroxy compounds. A few examples are collected in Scheme 19. The enzymatic resolution of substituted cyclohexanols **53** (cis and trans) is an attractive route to chiral auxiliaries and reagents like menthol or phenylmenthol [13]. Many derivatives, including menthol, have been successfully resolved using, for example, CCL-catalyzed esterification with lauric acid in heptane [153] or PSL-catalyzed transesterification with vinyl acetate in t-BuOMe [154]. A similar example is the resolution of glycal derivatives, which are valuable participants in the synthesis of glycoconjugates and oligosaccharides [155]. Bicyclo [2.2.1] hept-5-en-2-ol **54** is resolved with vinyl acetate and CCL, modified by selectively blocking its ε-amino residues of lysine with epoxy-activated polymer support. This immobilization leads to a five-fold increase of selectivity which is entirely preserved against deactivation caused by acetaldehyde, an unavoidable by-product in acyl transfer reactions with vinyl acetate [156]. Racemic **55** is resolved with vinyl acetate and PFL to afford both the ester and the remaining alcohol with ee > 95% [157]. The products are valuable precursors to various classes of carbocyclic nucleosides, particularly of the 2',3'-dideoxydidehydro type. The optically pure ester of **55** was converted in three steps to (+)-carbovir, a possible chemotherapeutic agent for the treatment of

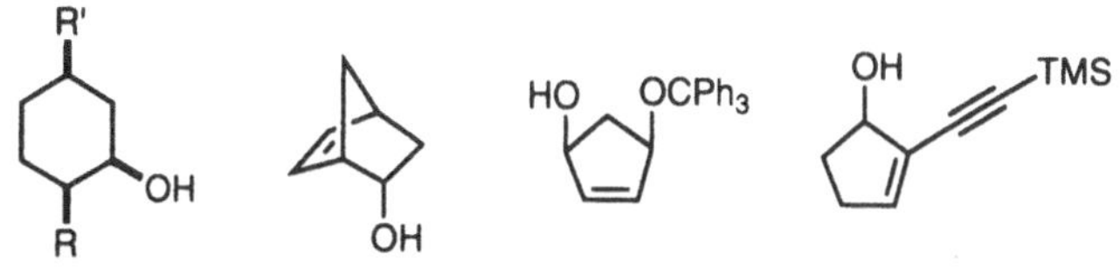

53
R=Ar, alkyl
R'=H, alkyl

54

55

56

Scheme 19

AIDS infections [157]. An interesting example is the resolution of racemic **56**. Racemic **56** is reacted with vinyl acetate in the presence of PSL in toluene affording ee $\geq$ 99 for the product and the starting material. Then, after removal of the insoluble enzyme and without further purification, the reaction mixture is subjected to the Mitsunobu reaction to convert the unreacted (−)-**56** into the corresponding (+)-acetate with inversion of configuration [158]. The (+)-acetate is reduced to the (+)-alcohol **56** (ee = 87.5%) and after one crystallization, (+)-**56** is obtained in an optically pure form (ee $\geq$ 99) and 73% overall yield from the racemic **56**. This work offers a solution to the problem of losing 50% of material in kinetic resolutions. Other examples of resolving cyclic secondary alcohols exist in the literature [159–161].

Trans-cyclodiols can also be subjected to enzymatic resolution, for instance, racemic 1,2:5,6-di-O-cyclohexylidene-myo-inositol **57** (Scheme 20), which is an important intermediate for synthesizing inositol phosphate derivatives. The diol **57** is treated with acetic anhydride and CCL in diethyl ether to yield almost equal amounts of the remaining diol and the corresponding monoacetate (at the C-4 position only), both with ee > 98% [162]. Enantiomerically pure *trans*-cycloalkane-1,2-diols **58** (Scheme 20) can be used for the synthesis of optically active crown ethers or as auxiliaries for the preparation of bidentate ligands. PSL-catalyzed acylation with vinyl acetate in *t*-BuOMe produces varying amounts of diesters, monoesters and unreacted diols with moderate to high optical purities, depending on the ring size [163]. The method is applicable to linear diols as well. PSL catalyzes the enantioselective diacetylation of racemic 2,3-butanediol **59** (Scheme 20) in vinyl acetate. Both acetylation steps favors the R-enantiomer; thus the reaction is a sequential kinetic resolution and the enantioselectivities of the two steps reinforce one another. The (S,S)-diol and the (R,R)-diacetate are obtained almost optically pure (ee > 96, 23-30%), while the (S,S)-monoacetate has low ee (21%) [164]. It has also been reported that mixtures of all stereoisomers (i.e R,R, S,S and R,S) of linear diols can be resolved successfully [165, 166].

Since many organometallic hydroxy compounds are decomposed by water, their resolution by asymmetric enzymatic hydrolysis is impossible, and therefore, the use of asymmetric esterification or transesterification in organic

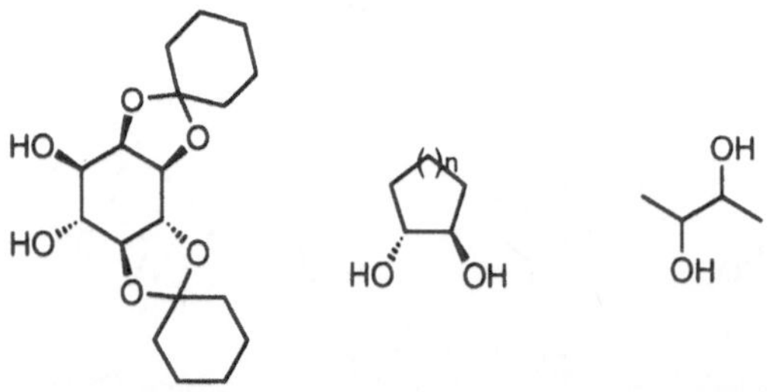

±57   ±58   ±59

Scheme 20

solvents is a necessity. Tricarbonyl chromium complexes constitute such a class of hydrolytically labile molecules and are easily resolved by enzyme-mediated transesterification [167, 168]. An example is the PSL-catalyzed enantioselective transesterification of racemic **60** with isopropenyl acetate (Scheme 21) [167]. The acetate **(S)-61** is obtained in 48% and 98% ee, and the remaining alcohol **(R)-60** is optically pure (100% ee, 47%). The resolution of oxime derivatives of chromium complexes has recently been reported [169]. Another example is the resolution of 1-ferrocenylethanol **62** (Scheme 22). Using PFL-catalyzed acyl-transfer with vinyl acetate in t-BuOMe, the acetate **(R)-63** is obtained with 96% ee at 50% conversion and the unreacted alcohol **(S)-62** has 92% ee [170]. If PPL is used for the same resolution with vinyl propanoate in toluene, the same stereochemistry is observed but with lower ee values (84% for both the R-acetate and the S-alcohol) [38].

Cyanohydrin compounds are versatile synthons in organic synthesis. They can be converted into α-hydroxycarboxylic acids, α-hydroxyaldehydes or ethanolamine derivatives. Several optically pure cyanohydrins have been prepared via oxynitrilase catalyzed addition of hydrogen cyanide to aldehydes in organic media [171], but only one enantiomer can be obtained in this way. Another method for their preparation involves the use of hydrolytic enzymes in the asymmetric hydrolysis of cyanohydrin acetates [172]. In this process, the unreacted ester is recovered in optically active form, but the cyanohydrin product is disregarded due to its spontaneous racemization in aqueous solution [173]. Optically active cyanohydrins as well as the corresponding esters can be obtained without racemization via enzymatic transesterification of racemic cyanohydrins in organic solvents. Using this method, several aryl cyanohydrins of synthetic value have been prepared with high optical purities [130, 173]. A novel approach to the synthesis of cyanohydrin acetates by in situ formation and racemization of cyanohydrins has recently been reported [174]. Racemic cyanohydrins **65** (Scheme 23), generated from aldehydes **64** and acetone cyanohydrin in diisopropyl ether under the catalysis of basic anion-exchange resin

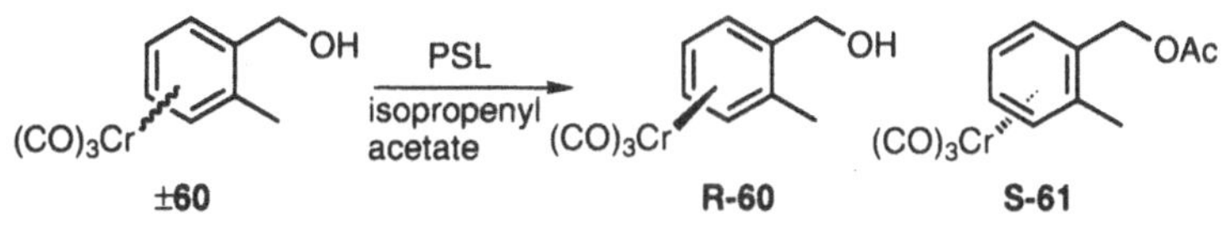

Scheme 21

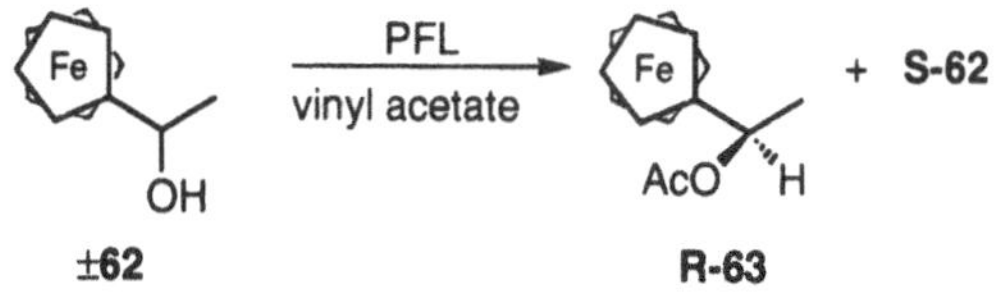

Scheme 22

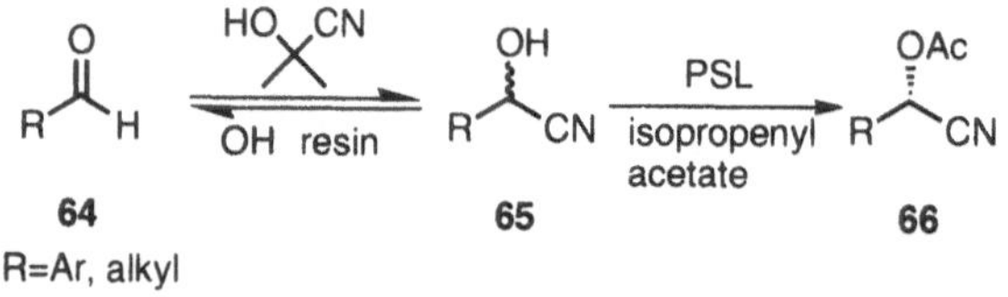

Scheme 23

(OH⁻ form), are stereoselectively acetylated by PSL with isopropenyl acetate as an acylating agent. The *S*-isomer of **65** is preferentially acetylated, while the unreacted *R*-isomer is continuously racemized through reversible transhydrocyanation catalyzed by the resin. The process enables one-stage conversion of various aldehydes **64** into the corresponding (*S*)-cyanohydrin acetates **66** with up to 94% ee in 63–100% conversion yields [174]. The in situ racemization of the substrate allows for quantitative conversion of racemic substrate to a single enantiomer of the product, thereby maximizing the chemical yield of the resolution to 100%. Many other interesting resolutions exist in the literature, for example the resolution of racemic binaphthol, that has an axial chirality [175, 176], or the ring opening of racemic epoxides with 2-propylamine to afford selectively (*S*)-propanol amines [177].

3.4 Resolution of Racemic Acids and Alcohols via Racemic Esters

Kinetic resolution may be carried out by transesterifying one enantiomer of a racemic ester, in which the asymmetric carbon is in the acid moiety, with an achiral alcohol. In this case both enantiomers are left in the form of esters, and separation may require a difficult chromatography or a careful distillation [178]. This is probably a major reason for the very limited application of the method. Several examples are collected in scheme 24. (2*S*,3*R*)-*iso*-Butyl *trans*-β-phenyl-glycidate **67a** (Scheme 24) is obtained from the corresponding racemic methyl ester by MML-mediated transesterification with *iso*-butanol in hexane (1:1, v/v [179]). At 36% and 45% conversion, the ee values for the product fraction are 97% and 95%, respectively; those for the remaining substrate fraction are 56% and 77%, respectively. Both the product and the remaining substrate can be utilized to prepare, in several chemical steps, the taxol C-13 side

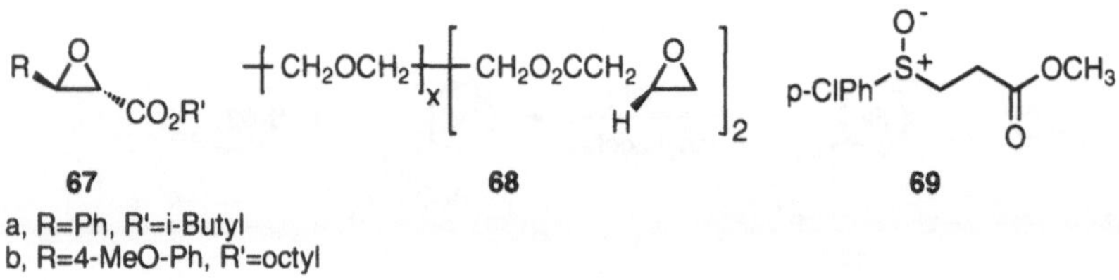

67

a, R=Ph, R'=i-Butyl
b, R=4-MeO-Ph, R'=octyl

68

69

Scheme 24

chain [179]. Taxol itself is an antimicrotubule agent isolated from the bark of *Taxos brevifolia* and has recently attracted much attention because of its efficacy in the treatment of various types of cancer. A similar example is compound **67b** (Scheme 24), a key intermediate in the synthesis of diltiazem which is one of the most potent calcium-channel blockers. Subjecting the racemic methyl ester to CCL catalyzed transesterification with octanol in *tert*-pentyl alcohol affords **(2S,3R)-67b** in 97% ee, but with low yield (maximum 20% [180]). Separation of the product from the remaining starting material can be achieved by flash chromatography with Et_2O:hexane (2:8, v/v).

In order to improve the ease of separating the two esters, it is possible to conduct the transesterification reaction with an alcohol that would impart very different properties to the modified enantiomer. It has been shown that racemic trichloroethyl 3,4-epoxybutanoate can be selectively transesterified by low molecular weight poly(ethylene glycol) (PEG) using PPL in diisopropylether [178]. At 50% conversion the PEG ester **(S)-68** is obtained with 89% ee and the remaining substrate has an ee of 96%. Separation of the PEG ester from the unreacted trichloroethyl ester is achieved by cooling the reaction mixture to 0 °C and filtering off the enzyme and the solidified **68**. The PEG can be removed in 92% yield by converting **68** to the corresponding methyl ester, using PPL-catalyzed transesterification with methanol. Transesterification is also used to prepare optically active sulfinylalkanoates which are compounds containing ester and sulfoxide functionalities and are useful reagents for organic synthesis. (S)-(−)-*n*-Butyl 3-[(4-chlorophenyl)sulfinyl]propanoate **69** is obtained by PSL-catalyzed transesterification of the racemic methyl ester with butanol in hexane, at 50% conversion the butyl ester is attained in > 95% ee and the remaining methyl ester has an ee of 90% [181]. The procedure is suitable for preparation of sulfinylalkanoates, where the ester and sulfoxide groups are separated by one or two methylene units, but compounds with three methylene groups are not substrates for the lipase.

Transesterification can also be used for resolution when the asymmetric carbon is in the alcohol moiety of the racemic ester. This process is particularly useful for *O*-alkyl esters of sterically hindered secondary alcohols because these sterically hindered alcohols hardly react by the direct transesterification methods. The racemic *O*-alkyl esters are subjected to lipase-catalyzed transesterification with achiral primary alcohols as nucleophiles and the desired chiral alcohol is released from the ester. The process is examplified in the resolution of racemic trifluoromethyl-1-(9-anthryl)ethyl butyrate **70**, the alcohol of which is an important chiral shift reagent and serves as key ingredient of stationary phases for chiral chromatography columns (Scheme 25 [182]). Racemic **70** is subjected to PPL catalysis in neat butanol to afford almost optically pure (R)-trifluoro-1-(9-anthryl)ethanol **71** [183]. The rate of the enzymatic reaction can be enhanced considerably by simple dispersion of the enzyme on neutral aluminium oxide. Under these conditions the reaction proceeds to 50% conversion, when all the (R)-ester has undergone the alcoholysis. The same approach is used for the resolution of the racemic ester **72** (Scheme 26). At 50% conversion,

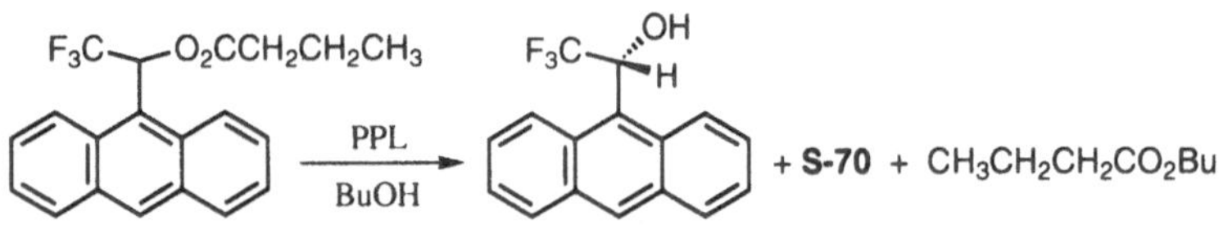

Scheme 25

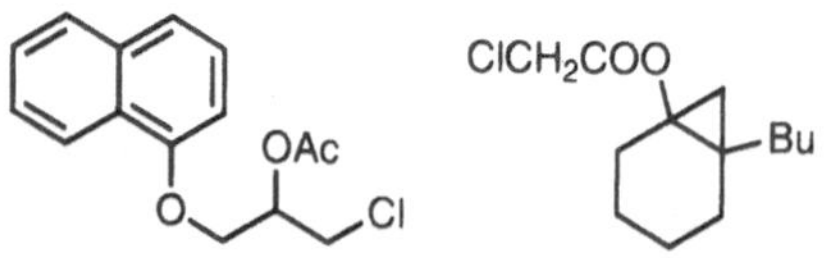

Scheme 26

transesterification of **72** with PSL and butanol in diisoproylether gives both the S-alcohol and the remaining R-acetate with ee > 95% (E > 100 [184]). The chiral alcohol and ester are then converted, in one step, to chiral propranolol, which is an important β-blocker, by treating with aqueous isopropylamine in the presence of NaOH. This procedure is used successfully for other resolutions of racemic esters [185, 186] and for the resolution of thioesters [187]. An exceptional example of resolving an ester of tertiary alcohol is the resolution of the cyclopropyl ester **73**. The transesterification is run in the presence of 1-propanol with MML in diisopropylether. After about 50% conversion (2.5–3 h) the (1S,6S)-alcohol is received with 86% ee while the remaining (1R,6R)-ester has an ee of 90% [188]. Other derivatives of cyclopropyl esters have also been checked [188].

If amines act as nucleophiles towards the intermediary acyl-enzyme, the synthesis of chiral amides from the racemic esters becomes possible. A simple application of this kind is the enantioselective synthesis of a number of (S)-amides from racemic alkyl 2-halopropanoates catalyzed by CCL (up to 95% ee) [189, 190]. Another interesting reaction is the asymmetric cleavage of oxazol-ones in organic solvents. In such an organic microenvironment, nonenzymatic hydrolysis proceeds very slowly but the rate of enolization of the C-4 proton is sufficiently rapid so that 100% of the substrate is convertible into product (Scheme 27). PSL is used to catalyze the enantioselective methanolysis of a

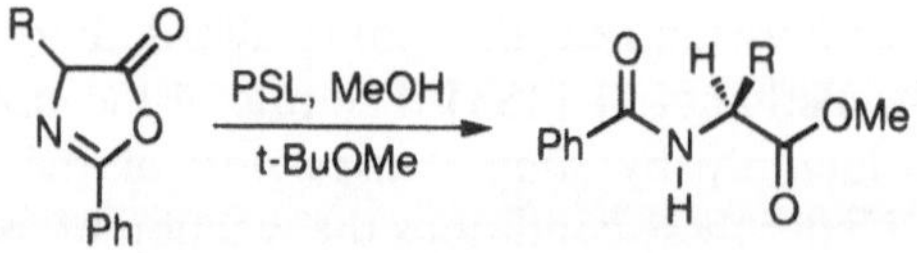

Scheme 27

variety of 4-substituted 2-phenyloxazolin-5-one derivatives **74** in *tert*-butyl-methylether to furnish optically active *N*-benzoyl-L-α-amino acid methyl esters **75** (31–93%, 66–98% ee) [191]. In earlier work, *n*-butanol was used in diisipropylether for the alcoholysis of 2-phenyl-4-methyloxazolin-5-one with MML but the optical purity of the product was only modest (57% ee) [192]. This approach is also used in the cleavage of β-lactam rings to yield derivatives of (2*R*,3*S*)-phenylisoserine, an important intermediate in the synthesis of the C-13 side chain of taxol, in high ee's [193], and also in the enantioselective cleavage of α-substituted cyclic acid anhydrides [194].

3.5 Resolution of Hydroxyesters – Intra and Intermolecular Transesterifications

Since hydroxyesters are bifunctional molecules, they undergo condensation by two alternative routes; intramolecular, to give the corresponding lactone, or intermolecular, to give oligomerization products. The outcome depends on the chain length and degree of substitution of the substrate [195] as well as on the reaction conditions. If a five membered lactone can be formed (from γ-hydroxyesters), this is the exclusive product [196]. For δ-hydroxyesters the outcome depends on substitution in position δ: if unsubstituted, oligomerization is predominant, while in the case of δ-methyl-γ-hydroxyester, the balance is tipped in favor of lactonization, and the corresponding δ-methylvalerolactone is the exclusive product [195]. The use of γ or δ-hydroxyacids instead of hydroxyesters as substrates for enzymatic resolution is not feasible, since hydroxyacids are prone to spontaneous lactonization. It is noteworthy that although chiral hydroxyesters may be obtained by enzymatic hydrolysis of the corresponding *O*-acyl derivatives, this approach is bound to fail because of undesirable hydrolysis at the carboxyl moiety.

The preparation of optically active γ-butyrolactones via enzymatic resolution of racemic γ-hydroxyesters is exemplified in Scheme 28. At low conversion rates (21–48%), PPL in diethyl ether yields the γ-butyrolactones **77** in high optical yield (ee = 82–98%). At high conversion rates (above 50%) the remaining hydroxyester **78** is obtained with high ee [60]. Other resolutions of γ-hydroxyesters [197, 198] as well as δ-hydroxyesters [199, 200] have been published, in all cases chiral lactones are obtained. If the hydroxy group is located in a more remote position, a chiral macrocyclic lactone of type **79** or **80** can be obtained (Scheme 29) [201–204]. In certain cases, tri-lactones (n = 8,14) or

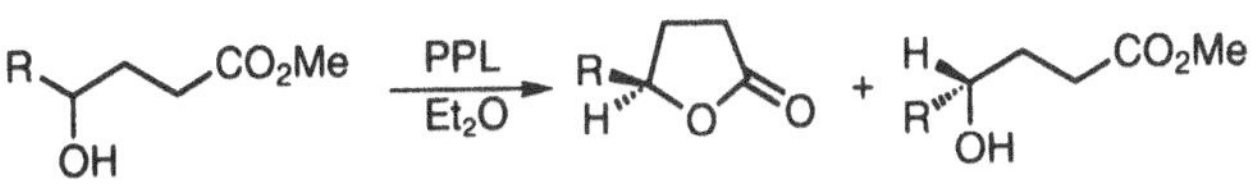

Scheme 28

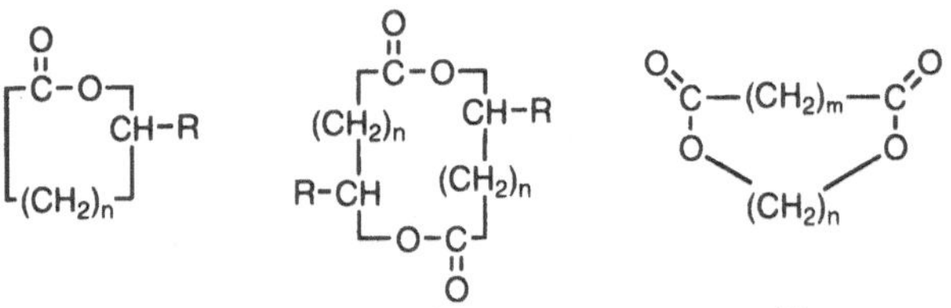

Scheme 29

tetra-lactones (n = 8) have been obtained in different ratios, depending on the lipase used [203].

Another approach to enzymatic polycondensation is lipase-catalyzed esterification of diacids with diols (or diesters and diols) in organic solvents. With α,ω-diacids (m = 2–12) and α,ω-diols (n = 5–16) the intermolecular transesterification, catalyzed by CCL or PSL in isooctane, leads to macrocyclic mono-lactones of type **81** (Scheme 29) or dilactones of that type [205]. As has already been mentioned, the alternative to lactonization is oligomerization. For example, optically active oligoesters (trimers and pentamers) are prepared from racemic diesters and achiral diols or vice versa by using various lipases as asymmetric catalysts [206], or the polymerization of 10-hydroxydecanoic acid with CCL in hexane to afford an achiral polymer with the average of 52 monomers [207]. The same procedure is applied to intramolecular cyclization of aminoesters for the formation of lactams, and to intermolecular reaction between diesters and diamines for the formation of macrocyclic bislactams [208].

4. Exploiting the Enzymes' Regioselectivity and Chemoselectivity

The selective modification of polyfunctional organic compounds is a difficult synthetic problem and often demands cumbersome multistep procedures. In order to overcome these synthetic tasks, it is possible to exploit the enzymes' regioselectivity i.e. their ability to select among several identical groups within the same molecule, or the enzymes' chemoselectivity i.e. their ability to select among groups with similar reactivity but of different chemical nature [19]. The above mentioned properties have been used, in particular, as protective steps in polyhydroxy compounds like carbohydrates. The enzymes' properties have also been exploited in order to separate geometrical isomers.

4.1 Enzymatic Acylation of Polyhydroxy Compounds

The regio and chemoselective potential of hydrolytic enzymes is applied to several classes of compounds. For example, this feature is used for the selective acylation of various 1,2- and 1,3-diols of type **82** and **83** (Scheme 30) [209–212]. While using PPL with different acylating reagents (activated and nonactivated esters as well as anhydrides), the primary hydroxy group of all substrates is almost exclusively esterified. On the other hand, the enantioselectivity displayed in this PPL-catalyzed reaction, when indicated, is very low [210, 211]. When 2-methyl-2,4-pentanediol is subjected to the enzymatic reaction, under similar conditions, no esterification occurs due to the absence of a primary hydroxyl group [210]. The method was used for the preparation of several aliphatic and aromatic primary esters of chloramphenicol **84** (Scheme 30), a natural antibiotic that can be administered as the 3-*O*-ester [213].

Although, in general, lipases catalyze the chemoselective acylation of aliphatic 1,2-diols with low enantioselectivity, it is possible in certain cases to overcome this disadvantage by a lipase-catalyzed sequential acylation of these substrates [214, 215], thus exploiting the enzymes' enantioselectivity as well as the chemoselectivity. For instance, in the presence of PSL and vinyl acetate, the (S)-enantiomers of 13 racemic aliphatic 1,2-diols are converted at a low rate into the primary (R)-monoacetates. The corresponding (R)-enantiomers are converted at a higher rate into the (S)-diacetates, with moderate to high optical purities (E = 12–100) [214]. When two primary hydroxy groups are present in the molecule, the primary group at the far end position of the asymmetric carbon is acylated almost exclusively [210, 216]. If heterobifunctional molecules, like amino alcohols, are subjected to the enzymatic acylation, the site of modification (OH or NH_2) can be controlled by the enzyme and the acylating agent [217].

Likewise, it is possible to exploit the enzymes' chemoselectivity toward substrates containing phenolic groups in addition to aliphatic hydroxy groups. It has been shown [216] that the primary group is always esterified in the presence of free phenolic OH (with CCL and butanol in hexane). When a secondary OH group is present in the molecule along with the phenol, reaction rates are much lower and different tendencies are observed depending on the relative position of the OH groups and the structure of the substrate [216]. The regioselectivity, together with the enantioselectivity of PSL, is used for the resolution of racemic-*trans*-sobrerol **85** (Scheme 31), a mucolytic drug. In spite of differences in the pharmacological activity between its (+) and (−)-form, it is still produced and marketed as a racemate [218]. Using PSL immobilized on Hyflo Super Cell and vinyl acetate in tert-amyl alcohol, it is possible to produce,

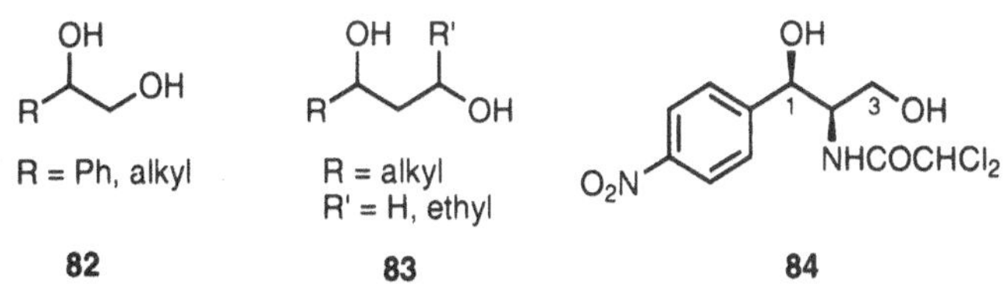

R = Ph, alkyl   R = alkyl
                       R' = H, ethyl

82   **83**   **84**

Scheme 30

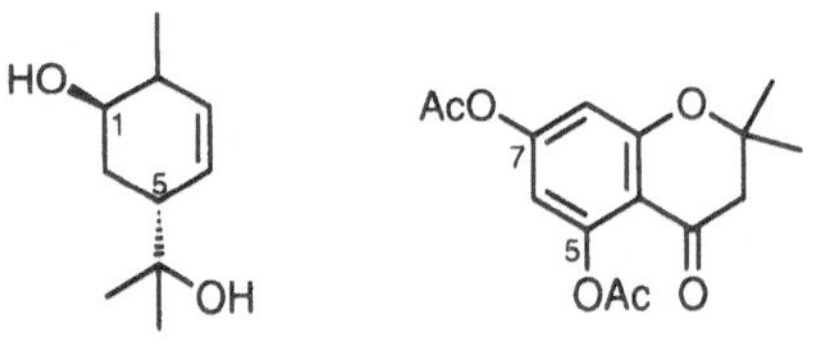

(1R,5S)-(+)-85   86

Scheme 31

at 50% conversion, both **(1R, 5S)-(+)-85** and the corresponding (−)-trans-sobrerol monoacetate (at the phenol position exclusively), with practically 100% optical purity [218]. If the substrate possesses two phenol groups, like aromatic dihydroxy aldehydes and ketones, the hydroxyl other than the one ortho to the carbonyl is selectively acylated, when vinyl acetate is used as acyl donor and PCL as catalyst [219]. Moreover, since the regiopreference in alcoholysis (or hydrolysis) of peracetates of polyacetoxyaryl aldehydes and ketones is the same as that observed in the lipase mediated acylation, the alcoholysis allows formation of partially acylated compounds different from those obtained by the direct acylation [219, 220]. The regioselective alcoholysis is also used in the case of benzopyranone derivatives that occur widely in nature, many of which possess a variety of biological activities, i.e. antitumour, antiviral, antibiotic and antifungal [221, 222]. For example, in the alcoholysis of 5,7-diacetoxy-2,2-dimethylchromanone **86** with PPL and *n*-butanol in THF-*i*-Pr$_2$O, only the 7-hydroxy group is deacetylated, affording 5-acetoxy-7-hydroxy-2,2-dimethylchromanone is 73% yield (Scheme 31) [222].

In carbohydrate chemistry, the problems of selective protection and deprotection are accentuated due to the presence of multiple hydroxyl functions of very similar reactivity [223]. Nevertheless enzymatic reactions in organic solvents have proven successful also in this field. It has been shown in several cases that the primary hydroxyl function of unprotected furanose and pyranose-type sugars is selectively acylated by using, for example, PPL and trichloroethyl carboxylates in pyridine [34], or protease N from *Bacillus subtilis* with isopropenyl acetate in a benzene:pyridine mixture (2:1, v/v) [38]. The chemoselectivity obtained is 70–100% with 38–100% conversion rates. A number of alkyl furanosides and pyranosides were also selectively acylated at the primary hydroxyl group [223, 224], allowing, for instance, the selective esterification of ethyl D-glucopyranoside **87** with C$_8$–C$_{18}$ fatty acids and CAL in yields of 85–95% of the 6-O-monoesters **88** (Scheme 32) [224]. It should be mentioned

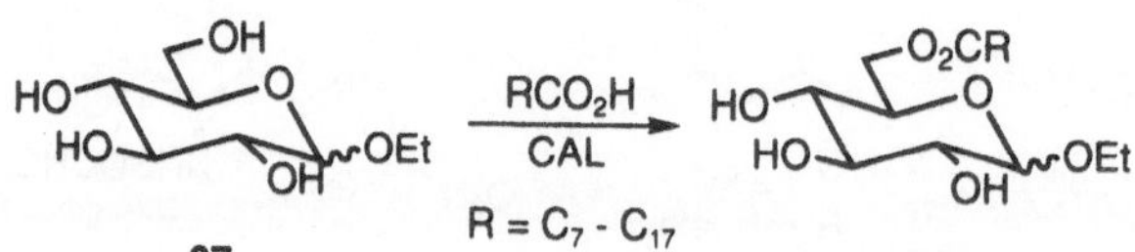

87                    $\xrightarrow[\text{CAL}]{\text{RCO}_2\text{H}}$ 88

R = C$_7$ - C$_{17}$

Scheme 32

that the fatty acids esters of carbohydrates constitute an interesting group of nonionic surface-active materials that exhibit highly useful properties [224]. If the monosaccharides are blocked at the primary hydroxyl group (enzymatically acylated or chemically alkylated), it is possible to exploit the enzymes' regioselectivity, thus enabling discrimination among the other available secondary hydroxyl functions [225–227]. While some lipases exclusively acylate the C-3 hydroxyl group, others display a preference toward the C-2 hydroxyl group [225]. An example is the CVL-catalysed acylation of 6-O-butyrylglucose 89 with trichloroethyl butyrate in THF (Scheme 33 [225]). The enzyme exhibits remarkable regioselectivity by acylating only the C-3 secondary hydroxyl group out of the four available, to afford 3,6-di-O-butyrylglucose (> 98% regioselectivity). In a certain case, 6-deoxy methyl α-L-pyranoside derivatives, like 90 (Scheme 33), can be acylated at the rather unreactive C-4 position by proper selection of the reaction conditions (PFL, THF:pyridine, 4:1, trifluoroethyl butyrate) [227]. The 4-butyryl ester of 90 is obtained in 45% yield and 97% regioselectivity.

The regioselective acylation methodology has been extended to carbohydrates other than monosaccharides, i,e, disaccharides and oligosaccharides. In these reactions the proteolytic enzyme subtilisin has been used mainly, since it was found to be catalytically active in dimethylformamide, the most suitable solvent for carbohydrates [228]. Several primary monobutanoyl esters of maltose 91 (Scheme 34), cellobiose, maltotriose and others have been synthesized on a preparative scale using trichloroethyl butyrate and subtilisin in dimethylformamide (45–57%, 75–95% regioselectivity) [228]. Interestingly, in the case of sucrose 92 (Scheme 34), acylation occurred regiospecifically on the 1'-O-hydroxy group, whereas in the chemical acylation the most reactive hydroxyls are the two primary ones at the C-6 and C-6' positions [228, 229]. Certain

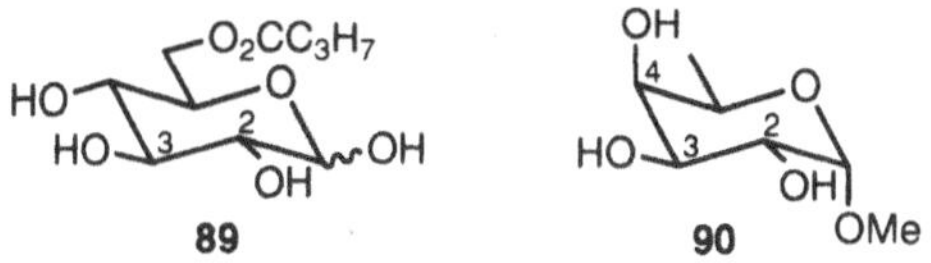

Scheme 33

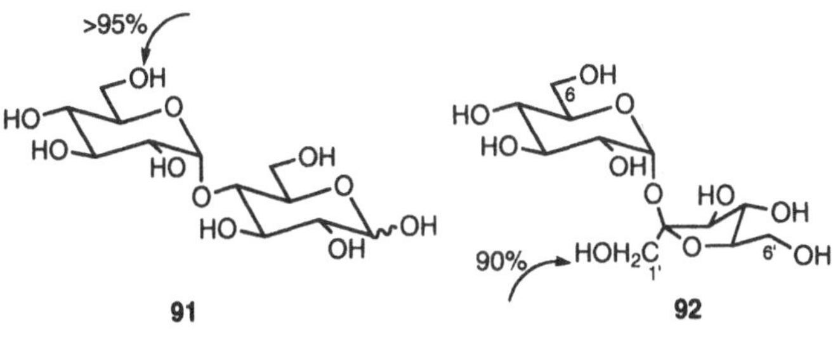

Scheme 34

flavonoid disaccharide monoglycosides have also been acylated by the catalytic action of subtilisin in pyridine [230]. The regioselectivity was high and the enzyme showed a preference for the C-3 position of the glucose moiety in disaccharides containing L-rhamnose and D-glucose.

The group of 4,6-O-benzylidene derivatives of various pyranosides, like compound 93 (Scheme 35), is acylated regiospecifically as well, in the C-2 or C-3 positions [231–233]. The position of acylation is dictated by the configuration at the anomeric centre. While α-derivatives are esterified predominantly at the C-2 OH, the β-derivatives are esterified chiefly at the C-3 hydroxy group. The reactions are conducted with PSL in either ethyl acetate [231, 233] or vinyl pivaloate (and other vinyl and trifluoroethyl carboxylates) in THF [232].

Nucleosides can also be subjected to enzymatic acylation in a regiospecific manner. Using subtilisin (or a modified subtilisin [234]) in dimethylformamide and trichloroethyl butyrate [228] or isopropenyl acetate [234], primary 5′-monoesters of various nucleosides and 2-deoxynucleosides were prepared with 55–100% regioselectivity. Likewise, the use of oxime derivatives as the acylating agents has been reported [235, 236]. It has been shown that PSL, in the presence of various oxime esters, catalyze the acylation of 2-deoxynucleosides at the 3′-position of the sugar moiety with total chemoselectivity, on the other hand, CAL has shown high chemoselectivity toward the primary hydroxyl group of both 2-deoxynucleosides and nucleosides.

Due to their polyfunctionalized rigid skeleton, steroids have always been a stimulating stereochemical exercise for organic chemists. Moreover, their phar-maceutical properties and their high manufacturing cost make any selective modifications of these natural compounds noteworthy [237]. It is possible to exploit the enzymes' regioselectivity and selectively acylate a given hydroxyl group in a steroid molecule. For instance the 5α-androstane-3β-17β-diol 94 (Scheme 36) can be enzymatically acylated with trifluoroethyl butyrate in

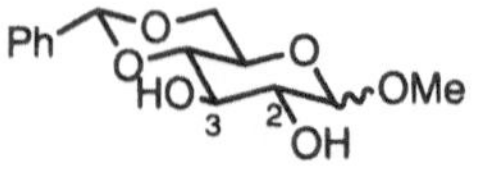

93

Scheme 35

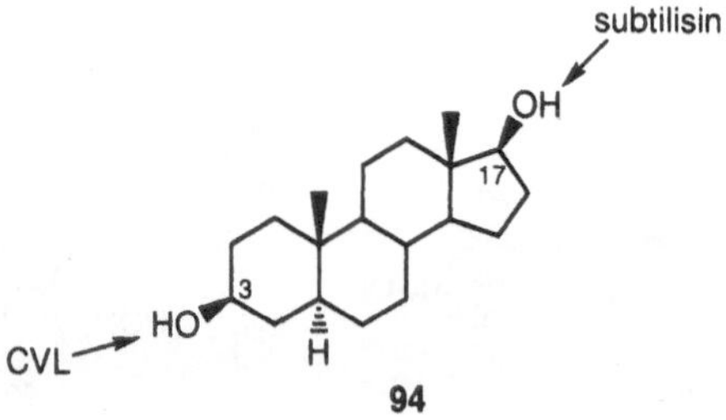

94

Scheme 36

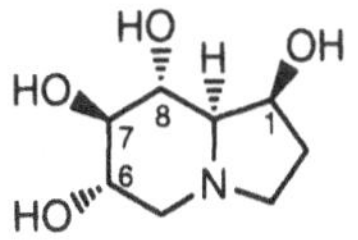

95

Scheme 37

anhydrous acetone [238]. CVL and subtilisin exhibit opposite regioselectivities in this reaction. While CVL reacts exclusively with OH at the C-3 position (83% yield), subtilisin displays a marked preference for the C-17 hydroxyl (60% yield). It should be mentioned that the chemical reactivities of these two hydroxyl groups are comparable. Other examples have been reported [237, 239, 240], including regioselective acylation of bile acid derivatives [237] and the deacylation (alcoholysis) of completely acylated polyhydroxy steroids [239].

The regioselective acylation of the alkaloid castanospermone **95** (Scheme 37) has been accomplished by action of subtilisin in pyridine [241]. The enzyme preferentially acylates the secondary OH group at the C-1 position. Various activated and enol esters of alkyl, aryl or α-aminoalkyl carboxylates are employed as the acylating agents to give the corresponding 1-*O*-acyl derivatives (31–91% yield). The 1-esters thus obtained are further utilized as substrates in a second enzymatic acylation in THF [241]. While subtilisin shows a strong preference for acylation of the OH group at the C-6 position, the lipases PPL and CVL prefer the OH group at C-7. The regioselectivities in the second enzymatic esterifications are only moderate. The regioselective acylation of another alkaloid 1-deoxynojirimycin has also been reported [242]. Many other examples that exploit the enzymes' chemo and regioselectivity exist in the literature including selective acylation of several glycal derivatives [243], acylation of myoinositol derivatives [244] and the acylation of *cis*-1,3-cyclopentanediol derivatives (substituted in the 4-position) [245]. The last two examples exploit the enzymes' enantioselectivity as well as regioselectivity.

4.2 Separation of E/Z Stereoisomers

Until now, little has been reported on the ability of hydrolytic enzymes to distinguish between geometric isomers and, in particular, between E/Z stereoisomers. In several studies that examined the PPL or PSL-catalyzed resolution of isomerically-pure secondary allylic alcohols [129, 145, 149, 246], it was noticed that the enzymes exerted a large difference in both reactivity and enantioselectivity between E and Z isomers, and the preferred isomer was always the E. This observation was used, in several cases, for the separation of mixtures of E/Z stereoisomers *via* enzymatic esterification in organic solvents, albeit with moderate selectivities. The results of the separation of E/Z-primary allylic alcohols are collected in Table 9 (Scheme 38).

Table 9. Enzyme-catalyzed acylation of E/Z-primary allylic alcohols

	Substrate	96^1 $E:Z$	time (min)	97^2 $E:Z$	Ref.
96a	 C_6H_{13} ⌁ OH	1:1	5–390	2.8–2.2:1	[247]
96b	 C_7H_{15} ⌁ OH	5.3:1	15–300	15–13:1	[247]
96c	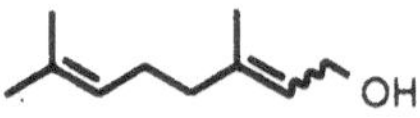 ⌁ OH	1:6.2	5–235	2.5–0.9:1	[247]
96c	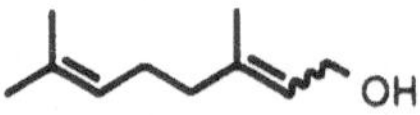 ⌁ OH	1:1	15–300	8.3–4:1	[248]

[1] The E/Z ratio of the starting alcohol.
[2] The E/Z ratio of the product

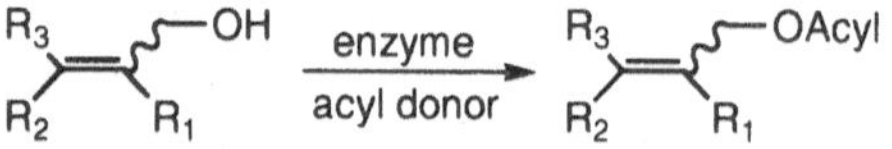

$$\begin{array}{ccc} R_3 & \text{—OH} & \xrightarrow[\text{acyl donor}]{\text{enzyme}} & R_3 & \text{—OAcyl} \\ R_2 & R_1 & & R_2 & R_1 \\ \textbf{96} & & & \textbf{97} \end{array}$$

Scheme 38

When an equimolar mixture of E and Z-2-nonenol **96a** was subjected to PPL-catalyzed acylation in Et$_2$O with a deficiency of trifluoroethyl butyrate, 2–3 time as much of the E-butyrate was formed over the course of the reaction (Table 9, scheme 38) [247]. In **96b** a methyl group was introduced in the 2-position, and while the $E:Z$ ratio in the starting material was 5.3:1, the ratio of the butyrate formed was approximately 14:1 (PPL, trifluoroethyl butyrate, Et$_2$O), again showing almost 3-fold preference for reaction with the E-isomer. The stereoisomeric mixtures of the allylic terpene alcohols E-geraniol and Z-nerol **96c**, used in flavor and fragrance preparations [20], were separated by PPL-catalyzed acylation in Et$_2$O with either trifluoroethyl butyrate [247], or hexanoic anhydride [248]. Once again, the E-geraniol was acylated faster to geranyl acetate, leaving the Z-nerol behind, although in one entry the starting $E:Z$ ratio was 1:6.2 (Table 9).

In the case of secondary allylic alcohols like 3-undecen-2-ol **98** and 4-phenyl-3-buten-2-ol **99** (Scheme 39), the E-isomer reacted 20–40 faster than the Z-isomer (PPL, trifluoroethyl butyrate, Et$_2$O [247]). Because of the high enantios-electivity of the enzymatic transesterification (E > 100 for the E-isomers), the S-enantiomer of the E-isomer did not participate in the reaction. The ability of lipases to differentiate between cis- and $trans$-1,4-cyclohexanedimethanol (**100, 101** respectively, Scheme 40) was investigated by transesterification of the commercially available diol mixture ($cis:trans$, ca. 1:2.5) with various lipases and acylating agents in t-BuOMe [249]. Best results were obtained when diethyl

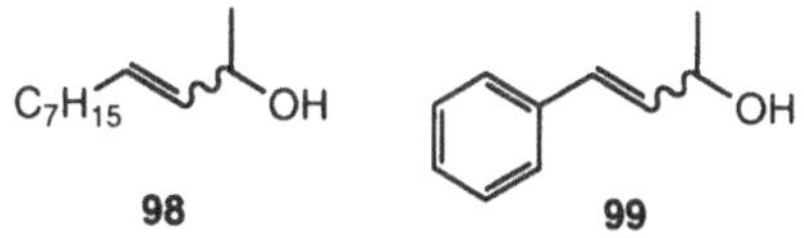

Scheme 39

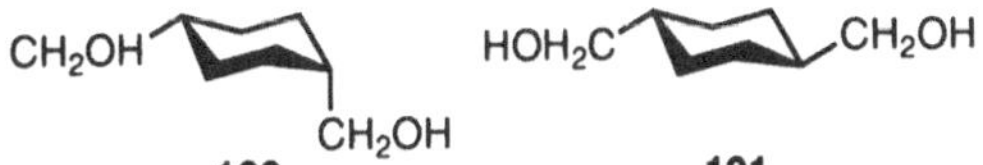

Scheme 40

fumarate was used as the acyl donor with PPL. The reaction was usually stopped at the mono- and di-transesterification, although oligomers were detected in all reactions. The monoester fraction showed a *cis:trans* mixture of 1:4.5, and the second transesterification yielded the *trans*-dicondensate with considerable selectivity- *cis:trans* ratio 1:24.

5 Effect of Solvent

An important advantage of working in organic solvents rather than in water is the possibility of altering the enzyme's properties by variations in the reaction medium. It has been shown that the nature of the solvent may have a profound effect on substrate specificity, as well as on the activity and stereoselectivity of enzymes. Water content is another important variable of the reaction medium, and its influence on enzyme activity and stereoselectivity has also been recently investigated. Many results, some of them contradictory, have been published, however, no rationale of general validity and predictive value has been presented so far, for the understanding of the effect of the solvent on the above mentioned properties. In this chapter some of these publications will be reviewed.

5.1 Effect on Reaction Rate and Substrate Specificity

The activity of various lipases and proteases can be markedly enhanced by alternations in the reaction medium. In several cases [250–254], the correlation between the reaction rate and the hydrophobicity of the solvent (expressed as log P) was observed, and the activity was shown to be greatest in the more hydrophobic solvents. For example, the initial rate of lipase- (PPL, CCL) catalysed transesterification reaction between tributyrin and heptanol was shown to be low in polar solvents (log P < 2), moderate in solvents having a log P between 2 and 4, and high in apolar solvents with a log P > 4 [250]. It was found that this correlation between polarity and activity parallels the ability of organic solvents to distort the essential water layer that stabilizes the bio-catalysts [250, 251]. Contrary to these publications, others [169, 255, 256] found no correlation with physicochemcial characteristics of the solvents such as hydrophobicity and dielectric constant, even though marked effect of the solvent was still observed.

The effect of the water content in the solvent on enzymatic activity was also investigated. In most cases, the enzymatic activity was greatly enhanced upon an increase in the water content in the solvents [257–260]. For instance, about 9-fold increase in activity was observed when 0.125% water were added to n-hexane in the CCL-catalysed esterification of α-bromopropanoic acid with n-butanol in hexane [258]. Water-mimicking solvents such as formamide and ethylene glycol, which can form multiple hydrogen bonds, can, at least partially, substitute for water as enzyme activators. The reaction rate of CCL-catalysed esterification of α-bromopropanoic acid with butanol in hexane was enhanced by 1.5–1.9 fold when formamide or ethylene glycol (0.05%) were added [258].

In the PPL-catalyzed transesterification of sulcatol (6-methyl-5-hepten-2-ol) with trifluoroethyl butyrate, the opposite phenomenon was observed: the addition of molecular sieves that remove water from the reaction medium markedly increased the transesterification rates, especially in the case of hydro-phobic solvents [256]. Thus, the rates increased about 4 times with toluene, cyclohexane and dodecane, whereas small increases or none were obtained with polar solvents as 3-pentanone and 3-methyl-3-pentanol. The addition of various additives to the organic solvents like carbohydrates [261], salt hydrates [262] or tertiary amines [263] was also studied and showed to have significant effect on reaction rates.

As was mentioned before, substrate speficicity can be tailored by changing the reaction medium as well [254, 264]. In an attempt to predict the solvent dependence of the enzyme's substrate specificity, a thermodynamic model was derived that correlated the substrate specificity with the solvent-to-water parti-tion coefficients of the substrates [264, 265]. Another approach for altering the substrate specificity is the water content of the medium. In the PPL-catalyzed transesterification between tributyrin and various alcohols, the dry lipase, in contrast to its wet counterpart, did not react with bulky tertiary alcohols [21].

5.2 Effect on Stereoselectivity

The enzyme's stereospecificity (i.e. prochiral selectivity, enantioselectivity and regioselectivity) can be profoundly affected simply by switching from one organic solvent to another. This phenomenon, although not quite understood at present, makes it possible to alter the specificity of a given enzyme at will and provides a valuable alternative to enzyme screening [266]. The dependence of the enzyme's prochiral selectivity on the solvent was observed in enzyme-catalyzed hydrolysis of prochiral diesters in hydrated organic solvents [55, 266] as well as in enzyme-catalyzed transesterification of prochiral diesters [259]. In the latter example [259], the transesterification of prochiral 2-benzyl-2-methyl dimethylmalonate with benzyl alcohol was investigated, and a 50% increase in LAPH-prochiral selectivity (expressed as the ee of the chiral product) was achieved simply by adding small quantities of water, with the optimum at 5% water in cyclohexane.

The majority of publications dealt with the effect of the nature of the solvent and its water content on the enantioselectivity of the enzymes. In a number of these publications, it was shown that the enantioselectivity of the enzyme decreased in order of the increase in the hydrophobicity of the solvent [26, 267], or increase in its dielectric constant and dipole moment [268]. Another study [252] showed a maximum of enantioselectivity at log P around 2 and lower enantioselectivities were obtained with solvents having higher or lower hydrophobicities. Similar behavior was observed in the PSL-catalyzed transesterification of 1-nitro-2-propanol. However, only when the solvents were divided into two groups, cyclic and acyclic, did both solvent groups afford smooth curves of E (enantiomeric ratio) vs log P [269]. On the other hand, in several cases no correlation was found with these physicochemcial properties [27, 253, 256]. It should be mentioned that in each investigation a different model reaction was studied (enzyme, substrate and acyl donor), making the comparison of results difficult. In several cases, the effect of the solvent was so pronounced, that inversion of enantioselectivity was observed upon changing from one solvent to another [169, 253, 267, 270].

The effect on enantioselectivity of water added to the organic solvent was studied as well. In certain works [126, 253, 268], the addition of water to the reaction medium lowered the enantioselectivity. For example, in the trans-esterification of sec-phenethyl alcohol with vinyl butyrate catalyzed by sub-tilisin, the addition of as little as 0.2% water to dioxane lowered the enzyme enantioselectivity from 50 to 18 and still further to 14 when the water content was doubled (the enantioselectivity was defined as the ratio of the initial rates of the S and R enantiomers) [268]. In contrast, in CCL-catalyzed esterification of α-bromopropanoic acid with butanol in hexane, the addition of water increased the E value from 17 (0% water) to 81 (0.125% water [258]). In one case, the alcoholysis of dibutyrylated-octylhydroquinone, the regioselectivity of PSL as a function of the solvent was investigated, and found to be reversed upon changing the reaction medium from toluene to acetonitrile [271]. In conclusion, the understanding of the effect of the solvent is essential, but no comprehensive model has been suggested to date that encompasses all of the existing results.

6. References

1. Midland MM, Nguyen NH (1981) J. Org. Chem. 46: 4107
2. Crossley R (1992) Tetrahedron 48: 8155
3. Jones JB, Sih CJ, Perlman D (eds) (1976) Applications of biochemical systems in organic chemistry. Wiley, New York

4. Porter R. Clark S (eds) (1985) Enzymes in organic synthesis. Ciba Foundation Symposium 111. Pitman, London
5. Davies HG, Green RH, Kelly DR, Roberts SM (eds) (1989) Biotransfromations in preparative organic chemistry. Academic Press, London
6. Abramowicz DA (ed) (1990) Biocatalysis. Van Nostrand Reinhold, New York.
7. Whitesides GM, Wong C-H (1985) Angew, Chem. Int. Ed. Engl. 24: 617.
8. Jones JB (1986) Tetrahedron 42: 3351.
9. Yamada H, Shimizu S (1988) Angew, Chem. Int. Ed. Engl. 27: 622
10. Wong C-H (1989) Science 244: 1145
11. Zhu L-M, Tedford MC (1990) Tetrahedron 46: 6587
12. Xie Z-F (1991) Tetrhedron Asymmetry 2: 733
13. Boland W, Frobl C, Lorenz M (1991) Synthesis 1049
14. Santaniello E, Ferraboschi P, Grisenti P, Manzocchi A (1992) Chem. Rev. 92: 1071
15. Faber K, Franssen MCR (1993) TIBTECH 11: 461
16. Chen C-S, Sih CJ (1989) Angew. Chem. Int. Ed. Engl. 28: 695
17. Dordick JS (1989) Enzyme Microb. Technol. 11: 194
18. Klibanov AM (1990) Acc. Chem. Res. 23: 114
19. Margolin AL (1991) CHEMTECH 160
20. Faber K, Riva S (1992) Synthesis 895
21. Zaks A, Klibanov AM (1984) Science 224: 1249
22. Aldercreutz P. Mattiasson B (1987) Biocatalysis 1: 99
23. Inagaki M, Hiratake J, nishioka T, Oda J (1991) J. Am. Chem. Soc. 113: 9360
24. Zaks A, Klibanov AM (1986) J. Am. Chem. Soc. 108: 2767
25. Russel AJ, Klibanov AM (1988) J. Biol. Chem. 263: 11624
26. Sakurai T, Margolin AL, Russell AJ, Klibanov AM (1988) J. Am. Chem. Soc. 110: 7236
27. Kitaguchi H, Fitzpatrick PA, Huber JE, Klibanov AM (1989) J. Am. Chem. Soc. 111: 3094
28. Klibanov AM (1986) CHEMTECH 16: 354
29. Huang FC, Lee LFH, Mittal RSD, Ravi Kumar PR, Chan JA, Sih CJ, Caspi E, Eck CR (1975) J. Am. Chem. Soc. 97: 4144
30. Ramos Tombo GM, Schar H-P, Fernandez X, Busquets I, Ghisalba O (1986) Tetrahedron Lett. 27: 5707
31. Janssen AJM, Klunder AJH, Zwanenburg BZ (1991) Tetrahedron 47: 7645
32. Zaks A, Klibanov AM (1985) Proc. Natl. Acad., Sci. USA 82: 3192
33. Chen C-S, Fujimoto Y, Girdaukas G, Sih CJ (1982) J. Am. Chem. Soc. 104: 7294
34. Therisod M, Klibanov AM (1986) J. Am. Chem. Soc. 108: 5638
35. Ghogare A, Kumar GS (1989) J. Chem. Soc., Chem. Commun. 1533
36. Bianchi D, Cesti P, Battistel E (1988) J. Org. Chem. 53: 5531
37. Sweers HM, Wong CH (1986) J. Am. Chem. Soc. 108: 6421
38. Wang Y-F, Lalonde JJ, Momongan M, Bergbreiter DE, Wong C-H (1988) J. Am. Chem. Soc. 110: 7200
39. Drueckhammer DG, Hennen WJ, Pederson RL, Barbas CF, Gautheron CM, Krach T, Wong C-H (1991) Synthesis 499
40. Baer E, Maurukas J, Russel MJ (1952) J. Am. Chem. Soc. 74: 152
41. Suemune H, Mizuhara Y, Akita H, Sakai K (1986) Chem. Pharm. Bull. 34: 3440
42. Hirth G, Barner R (1982) Helv. Chim. Acta 65: 1059
43. Chandrakumar NS, Hajdu J (1983) J. Org. Chem. 48: 1197
44. Morishima H, Koike Y, Nakano M, Atsuumi S, Tanaka S, Funabashi H, Hashimoto J, Sawasaki Y, Mino N, Nakano K, Matsushima K, Nakamichi K, Yano M (1989) Biochem. Biophys. Res. Commun. 159: 999
45. Kojke K, Numata M, Sugikoto M, Nakahara Y, Ogawa T (1986) Carbohydr. Res. 113
46. Tsuji K, Terao Y, Achiwa K (1989) Tetrahedron Lett. 30: 6189
47. Atsuumi S, Nakano M, Koike Y, Tanaka S, Ohkubo M, Yonezawa T, Funabashi H, Hashimoto J, Morishima H (1990) Tetrahedron Lett. 31: 1601
48. Bianchi D. Cesti P, Golini P, Spezia S, Garavaglia C, Mirenna L, (1992) Pure Appl. Chem. 64: 1073
49. Barnett CJ, Wilson TM (1989) Tetrahedron Lett. 30: 6291
50. Wang, Y-F, Wong C-H (1988) J. Org. Chem. 53: 3127
51. Terao Y, Murata M, Achiwa K, Nishio T, Akamtsu M, Kamimura M (1988) Tetrahedron Lett. 29: 5173
52. Santaniello E, Feraboschi P, Grisenti P, (1990) Tetrahedron Lett. 31: 5657

53. Djerourou A-H, Blanco L (1991) Tetrahedron Lett. 32: 6325
54. Naemura K, Furutani A (1991) J. Chem. Soc., Perkin Trans. 1: 2891
55. Hirose Y, Kariya K, Sasaki I, Kurono I, Ebiike H, Achiwa K, (1992) Tetrahedron Lett. 33: 7157
56. Holdgrun XK, Sih CJ (1991) Tetrahedron Lett. 32: 3465
57. Yamamoto K, Nishioka T, Oda J, Yamamoto Y (1988) Tetrahedron Lett. 29: 1717
58. Yamamoto Y, Iwasa M, Sawada S, Oda J (1990) J. Agric. Biol. Chem. 54: 3269
59. Murata M, Achiwa K (1991) Tetrahedron Lett. 32:6763
60. Gutman AL, Zuobi K, Bravdo T (1990) J. Org. Chem. 55: 3546
61. Gutman AL, Bravdo T (1989) J. Org. Chem. 54: 4263
62. Gao J, Jorgensen WL (1990) Chemtracts–Organic Chemistry 3: 244
63. Bjorkling F, Boutelje J, Gatenbeck S, Hult K, Norin T, Szmulik P (1985) Tetrahedron 41: 1347
64. Luyten M, Muller S, Herzog B, Keese R (1987) Helv. Chim. Acta 70: 1250
65. Toone EJ, Jones JB (1991) Tetrahedron Asymmetry 2: 1041
66. Gutman AL, Shapira M, Boltanski A (1992) J. Org. Chem. 57: 1063
67. Shapira M, Gutman AL (1994) Tetrahedron Asymmetry 5: 1689
68. Gutman AL, Bravdo T (1989) J. Org. Chem. 54: 5645
69. Theil F, Schick H, Winter G, Reck G (1991) Tetrahedron 47: 7569
70. Harris KJ, Gu Q-M, Shih Y-E, Girdaukas G, Sih CJ (1991) Tetrahedron Lett. 32: 3941
71. Johnson CR, Bis SJ (1992) Tetrahedron Lett. 33: 7287
72. Johnosn CR, Golebiowski A, McGill TK, Steensma DH (1991) Tetrahedron Lett. 32: 2597
73. Ader U, Breitgoff D, Klein P, Laumen KE, Schneider MP (1989) Tetrahedron Lett. 30: 1793
74. Bonini C, Racioppi R, Viggiani L, Righi G, Rossi L (1993) Tetrahedron Asymmetry, 4: 793
75. Bonini C, Racioppi R, Righi G, Viggiani L (1993) J. Org. Chem. 58: 802
76. Pottie M, Van der Eycken J, Vandewalle M (1991) Tetrahedron Asymmetry 2: 239
77. Tanaka M, Yoshioka M, Sakai K (1993) Tetrahedron Asymmetry 4: 981
78. Toyooka N, Nishino A, Mamose T (1993) Tetrahedron Lett. 34: 4539
79. Takano S, Moriya M, Higashi Y, Ogasawara K (1993) J. Chem. Soc., Chem. Commun. 177
80. Andreu C, Macro JA, Asensio G (1990) J. Chem. Soc., Perkin Trans. 1: 3209
81. Theil F, Ballschuh S, Schick H, Haupt M, Hafner B, Schwarz S (1988) Synthesis 540
82. Sugai T, Mori K (1988) Synthesis 19
83. Theil F, Kunath A, Schick H (1992) 33: 3457
84. Hemmerle H, Gais HJ (1987) Tetrahedron Lett. 28: 3471
85. Momose T, Toyooka N, Jin M (1992) Tetrahedron Lett. 33: 5389
86. Vanttinen E, Kanerva LT (1992) Tetrahedron Asymmetry 3: 1529
87. Sato M, Ohuchi H, Abe Y, Kaneko C (1992) Tetrahedron Asymmetry 3: 313
88. Bis SJ, Whitaker DT, Johnson CR (1993) Tetrahedron Asymmetry 4: 875
89. Nicolosi G, Morrone R, Patti A, Piattelli M (1992) Tetrahedron Asymmetry 3: 753
90. Howell JAS, Palin MG, Jaouen G, Top S, El Hafa H, Cense JM (1993) Tetrahedron Asymmetry 4: 1241
91. Fuji K, Kawabata T, Kiryu Y, Sugiura Y (1990) Tetrahedron Lett. 31: 6663
92. Ozegowski R, Kunath A, Schick H (1993) Tetrahedron Asymmetry 4: 695
93. Kagan HB (1985) Chiral ligands for asymmetric catalysis. In: Morrison JD (ed) Asymmetric synthesis, Vol 5. Academic, Orlando, p1
94. Gotor V, Memendez E, Mouloungui Z, Gaset A (1993) J. Chem. Soc., Perkin Trans. 1: 2453
95. Pozo M, Gotor V (1993) Tetrahedron 49: 4321
96. Puertas S, Brieva R, Rebolledo F, Gotor V (1993) Tetrahedron 49: 4007
97. Margolin AL, Fitzpatrick PA, Dubin PL, Klibanov AM (1991) J. Am. Chem. Soc. 113: 4693
98. Gutman AL, Meyer E, Kalerin E, Polyak F, Sterling J (1992) Biotechnol. Bioeng. 40: 760
99. Asensio G, Andreu C, Macro JA (1991) Tetrahedron Lett 32: 4197
100. Fernandez S, Brieva R, Rebolledo F, Gotor V (1992) J. Chem. Soc., Perkin Trans. 1 2885
101. Bevinakatti HS, Newadkar R (1990) Tetrahedron Asymmetry 1: 583
102. Cambou B, Klibanov AM (1985) Biotechnol. Bioeng. 26: 1449
103. Kirchner G, Scollar MP, Klibanov AM (1985) J. Am. Chem. Soc. 107: 7072
104. Engel KH (1991) Tetrahedron Asymmetry 2: 165
105. Berglund P, Holmquist M, Hedenstrom E, Hult K, Hogberg HE (1993) Tetrahedron Asymmetry 4: 1869
106. Chen C-S, Wu S-H, Girdaukas G, Sih CJ (!987) J. Am. Chem. Soc. 109: 2812
107. Fukui T, Kawamoto T, Sonomoto K, Tanaka A (1990) Appl. Microbiol. Biotechnol. 34: 330
108. Macfarlane ELA, Roberts SM, Steukers VGR, Taylor RL (1993) J. Chem. Soc., Perkin Trans. 1 2287

109. Delinck DL, Margolin AL (1990) Tetrahedron Lett. 31: 6797
110. Ferraboschi P, Grisenti P, Manzocchi A, Santaniello E (1990) J. Org. Chem. 55: 6241
111. Grisenti P, Ferraboschi P, Manzocchi A, Santaniello E (1992) Tetrahedron 48: 3827
112. Lentz N, Peet N (1990) Tetrahedron Lett. 31: 811
113. Bevinakatti HS, Newadkar RV (1993) Tetrahedron Asymmetry 4:773
114. Kanerva LT, Vanttinen E (1993) Tetrahedron Asymmetry 4: 85
115. Bianchi D, Cabri W, Cesti P, Francalanci F, Rama F (1988) Tetrahedron Lett. 29: 2455
116. Ferraboschi P, Casati S, Grisenti P, Santaniello E (1993) Tetrahedron Asymmetry 4: 9
117. Jouglet B, Rousseau G (1993) Tetrahedron Lett. 34: 2307
118. (a) Ennis MD, Old DW (1992) Tetrahedron Lett. 33: 6283 (b) Ennis MD, Ghazal NB (1992) Tetrahedron Lett. 33: 6287
119. Antus S, Gottsegen A, Kajtar J, Kovacs T, Toth TS, Wagner H (1993) Tetrahedron Asymmetry 4: 339
120. Herradon B (1992) Tetrahedron Asymmetry 3: 209
121. Janssen AJM, Klunder AJH, Zwanenburg B (1991) Tetrahedron 47: 5513
122. Morgan B, Oehlschlager AC, Stokes TM (1991) Tetrahedron 47: 1611
123. Frykman H, Ohrner N, Norin T, Hult K (1993) Tetrahedron Lett. 34: 1367
124. Hult K, Norin T (1992) Pure Appl. Chem. 64: 1129
125. Belan A, Bolte J, Fauve A, Gourcy JG, Veschambre H (1987) J. Org. Chem. 52: 256
126. Stokes TM, Oehlschlager AC (1987) Tetrahedron Lett. 28: 2091
127. Gutman AL, Brenner D, Boltanski A (1993) Tetrahedron Asymmetry 4: 839
128. Laumen K, Brietgoff D, Schneider MP (1988) J. Chem. Soc., Chem. Commun. 1459
129. Morgan B, Oehlschlager AC, Stokes TM (1992) J. Org. Chem. 57: 3231
130. Hsu S-H, Wu S-S, Wang Y-F, Wong C-H (1990) Tetrahedron Lett. 31: 6403
131. Nieduzak TR, Margolin AL (1991) Tetrahedron Asymmetry 2:113
132. Izumi T, Fukaya K (1993) Bull. Chem. Soc. Jpn. 66: 1216
133. Hiratake J, Inagaki M, Nishioka T, Oda (1988) J. Org. Chem. 53: 6130
134. Ader A, Schneider MP (1992) Tetrahedron Asymmetry 3:521
135. Bevinakatti HS, Banjeri AA (1992) J. Org. Chem. 57: 6003
136. Kim M-J, Choi YK (1992) J. Org. Chem. 57: 1605
137. Chen C-S, Liu Y-C, Marsella MJ (1990) J. Chem. Soc., Perkin 1 2559
138. Sakaki J-I, Sakoda H, Sugita Y, Sato M, Kaneko C (1991) Tetrahedron Asymmetry 2:343
139. Georgens U, Schneider MP (1991) J. Chem. Soc., Chem. Commun. 1064
140. Georgens U, Schneider MP (1991) J. Chem. Soc., Chem. Commun. 1066
141. Chong JM, Mar EK (1991) Tetrahedron Lett. 32: 5683
142. Katsuki T, Sharpless KB (1980) J. Am. Chem. Soc. 102: 5974
143. Carlier PR, Mungall WS, Schroder G, Sharpless KB (1988) J. Am. Chem. Soc. 110: 2978
144. Burgess K, Jennings LD (1990) J. Am. Chem. Soc. 112: 7434
145. Burgess K, Jennings LD (1991) J. Am. Chem. Soc. 113: 6129
146. Burgess K, Jennings LD (1990) J. Org. Chem. 55: 1138
147. Bornscheuer U, Schapohler S, Scheper T, Schugerl K (1991) Tetrahedron Asymmetry 2: 1011
148. Burgess K, Henderson I (1990) Tetrahedron Asymmetry 1: 57
149. Carretero JC, Dominguez E (1992) J. Org. Chem. 57: 3867
150. Carretero JC, Dominguez E (1993) J. Org. Chem. 58: 1596
151. Sparks MA, Panek JS (1991) Tetrahedron Lett. 32: 4085
152. Gil G, Ferre E, Meou A, Petit JL, Triantaphylides C (!987) Tetrahedron Lett. 28: 1647
153. Langrand G, Secchi M, Buono G, Baratti J, Triantaphylides C (1985) Tetrahedron Lett. 26: 1857
154. Laumen K, Seemayer R, Schneider MP (1990) J. Chem. Soc., Chem. Commun. 49
155. Berkowitz DB, Danishefsky SJ (1991) Tetrahedron Lett. 32:5497
156. Berger B, Faber K (1991) J. Chem. Soc., Chem. Commun. 1198
157. Evans CT, Roberts SM, Shoberu KA, Sutherland AG (1992) J. Chem. Soc., Perkin Trans. 1 589
158. Takano S, Suzuki M, Ogasawara K (1993) Tetrahedron Asymmetry 4: 1043
159. McCague R, Olive HF, Roberts SM (1993) Tetrahedron Lett. 34: 3785
160. Cotterill IC, Sutherland AG, Roberts SM, Grobbauer R, Spreitz J, Faber K (1991) J. Chem. Soc., Perkin Trans. 1 1365
161. Cregge RJ, Wagner ER, Freedman J, Margolin AL (1990) J. Org. Chem. 55: 4237
162. Ling L, Watanabe Y, Akiyama T, Ozaki S (1992) Tetrahedron Lett. 33: 1911
163. Seemayer R, Schneider MP (1991) J. Chem. Soc., Chem. Commun. 49
164. Caron G, Kazlauskas RJ (1993) Tetrahedron Asymmetry 4: 1995

165. Mattson A, Ohrner N, Hult K, Norin T (1993) Tetrahedron Asymmetry 4: 925
166. Bisht KS, Parmar VS, Crout DHG (1993) Tetrahedron Asymmetry 4: 957
167. Nakamura K, Ishihara K, Ohno A, Uemura M, Nishimura H, Hayashi Y (1990) Tetrahedron Lett. 31: 3603
168. Yamazaki Y, Hosono K (1990) Tetrahedron Lett. 31: 3895
169. Baldoli C, Maiorana S, Carrea G, Riva S (1993) Tetrahedron Asymmetry 4: 767
170. Boaz NW (1989) Tetrahedron Lett. 30: 2061
171. Effenberger F, Ziegler T, Forster S (1987) Angew. Chem. Int. Ed. Engl. 26: 458
172. Ohta H, Miyamae Y, Tsuchihashi G (1986) J. Agric. Biol. Chem. 50: 3181
173. Wang Y-F, Chen S-T, Liu KK-C, Wong C-H (1989) Tetrahedron Lett. 30: 1917
174. Inagaki M, Hiratake J, Nishioka T, Oda J (1992) J. Org. Chem. 57: 5643
175. Inagaki M, Hiratake J. Nishioka T, Oda J (1989) J. Agric. Biol. Chem. 53: 1879
176. Lin G, Liu S-H, Chen S-J, Wu F-C, Sun H-L (1993) Tetrahedron Lett. 34: 6057
177. Kamal H, Damayanthi Y, Rao MV (1992) Tetrahedron Asymmetry 3: 1361
178. Wallase JS, Reda KB, Williams ME, Morrow CJ (1990) J. Org. Chem. 55: 3544
179. Gou D-M, Liu Y-C, Chen C-S (1993) J. Org. Chem. 58: 1287
180. Kanerva LT, Sundholm O (1993) J. Chem. Soc., Perkin Trans. 1 1385
181. Burgess K, Henderson I, Ho K-K (1992) J. Org. Chem. 57: 1290
182. Pirkle WH, Sikkenga DL, Pavlin MS (1989) Chem. Rev. 89: 347
183. Shkolnik E, Gutman AL (1994) Bioorg. & Medicinal Chem. 2: 567
184. Bevinakatti HS, Banjeri AA (1991) J. Org. Chem. 56: 5372
185. Lu Y, Miet C, Kunesch N, Poisson JE (1993) Tetrahedron Asymmetry 4: 893
186. Bevinakatti HS, Banjeri AA, Newadkar RV (1989) J. Org. Chem. 54: 2453
187. Bianchi D, Cesti P (1990) J. Org. Chem. 55: 5657
188. Barnier JP, Blanco L, Rousseau G, Guibe-Jampel E, Fressel (1993) J. Org. Chem. 58: 1570
189. Quiros M, Sanchez VM, Brieva R, Rebolledo F, Gotor V (1993) Tetrahedron Asymmetry 4: 1105
190. Gotor V, Brieva R, Gonzalez, Rebolledo F (1991) Tetrahedron 47: 9207
191. Crich JZ, Brieva R, Marquart P, Gu R-L, Flemming S, Sih CJ (1993) J. Org. Chem. 58: 3252
192. Bevinakatti HS, Newadkar RV, Banjeri AA (1990) J. Chem., Soc., Chem. Commun. 1091
193. Brieva R, Crich JZ, Sih CJ (1993) J. Org. Chem. 58: 1068
194. Hiratake J, Yamamoto K, Yamamoto Y, Oda J (1989) Tetrahedron Lett. 30: 1555
195. Gutman AL, Oren D, Boltanski A, Bravdo T (1987) Tetrahedron Lett. 28: 3861
196. Gutman AL, Zuobi K, Boltanski A (1987) Tetrahedron Lett. 28: 5367
197. Huffer M, Schreier P (1991) Tetrahedron Asymmetry 2: 1157
198. Sugai T, Ohsawa S, Yamada H, Ohta H (1990) Synthesis 1112
199. Bonini C, Pucci P, Racioppi R, Viggiani L (1992) Tetrahedron Asymmetry 3: 29
200. Henkel B, Kunath A, Schick H (1993) Tetrahedron Asymmetry 4: 153
201. Makita A, Nihira T, Yamada Y (1987) Tetrahedron Lett. 28: 805
202. Yamada H, Ohsawa S, Sugai T, Ohta H, Yoshikawa S (1989) Chem. Lett. 1775
203. Zhi-wei G, Ngooi TK, Scilimati A, Fulling G, Sih CJ (1988) Tetrahedron Lett. 29: 5583
204. Lobell M, Schneider MP (1993) Tetrahedron Asymmetry 4: 1027
205. Zhi-wei G, Sih CJ (1988) J. Am. Chem. Soc. 110: 1999
206. Margolin AL, Crenne J-Y, Klibanov AM (1987) Tetrahedron Lett.. 28: 1607
207. O'Hagan D, Zaidi NA (1993) J. Chem. Soc., Perkin Trans. 1 2389
208. Gutman AL, Meyer E, Yue X, Abell C (1992) Tetrahedron Lett. 33: 3943
209. Cesti P, Zaks A, Klibanov AM (1985) Appl. Biochem. Biotechnol 11: 401
210. Parmar VS, Sinha R, Bisht KS, Gupta S, Prasad AK, Taneja P (1993) Tetrahedron 49: 4107
211. Janssen AJM, Klunder AJH, Zwanenburg B (1991) Tetrahedron 47: 7409
212. Ramaswamy S, Morgan B, Oehlschlager AC (1990) Tetrahedron Lett. 31: 3405
213. Ottolina G, Carrea G, Riva S (1990) J. Org. Chem. 55: 2366
214. Theil F, Weidner J, Ballschuh S, Kunath A, Schick H (1993) Tetrahedron Lett. 34: 305
215. Wang Y-F, Dumas DP, Wong C-H (1993) Tetrahedron Lett. 34: 403
216. Pedrocchi-Fantoni G, Servi S (1992) J. Chem. Soc., Perkin Trans 1 1029
217. Chinsky N, Margolin AL, Klibanov AM (1989) J. Am. Chem. Soc. 111: 386
218. Bovara R, Carrea G, Ferrara L, Riva S (1991) Tetrahedron Asymmetry 2: 931
219. Nicolosi G, Piattelli M, Sanfilippo C (1993) Tetrahedron 49: 3143
220. Parmar VS, Prasad AK, Sharma NK, Singh SK, Pati HN, Gupta S (1992) Tetrahedron 48: 6495

221. Natoli M, Nicolosi G, Piattelli M (1992) J. Org. Chem. 57: 5776
222. Parmar VS, Prasad AK, Sharma NK, Vardhan A, Pati HN, Sharma SK, Bisht KS (1993) J. Chem. Soc., Chem. Commun. 27
223. Hennen WJ, Sweers HM, Wang Y-F, Wong C-H (1988) J. Org. Chem. 53: 4939
224. Bjorkling F, Godtfredsen SE, Kirk O (1989) J. Chem. Soc., Chem. commun. 934
225. Therisod M, Klibanov AM (1987) J. Am. Chem. Soc. 109: 3977
226. Nicotra F, Riva S, Secundo F, Zucchelli L (1989) Tetrahedron Lett. 30: 1703
227. Ciuffreda P, Colombo D, Ronchetti F, Toma L (1990) J. Org. Chem. 55: 4187
228. Riva S, Chopineau J, Kieboom APG, Klibanov AM (1988) J. Am. Chem. Soc. 110: 584
229. Carrea G, Riva S, Secundo F, Danieli B (1989) J. Chem. Soc., Perkin Trans. 1: 1057
230. Danieli B, De Bellis P, Carrea G, Riva S (1990) Helv. Chim. Acta 73: 1837
231. Chinn MJ, Lacazio G, Spackman DG, Turner NJ, Roberts SM (1992) J. Chem. Soc., Perkin Trans. 1: 661
232. Panza L, Brasca S, Riva S, Russo G (1993) Tetrahedron Asymmetry 4: 931
233. Lacazio G, Roberts SM (1993) J. Chem. Soc., Perkin Trans. 1: 1099
234. Wong C-H, Chen S-T, Hennen WJ, Bibbs JA, Wang Y-F, Liu JL-C, Pantoliano MW, Whitlow M, Bryan PN (1990) J. Am. Chem. Soc. 112: 945
235. Moris F, Gotor V (1993) J. Org. Chem. 58: 653
236. Garcia-Alles LF, Moris F, Gotor V (1993) Tetrahedron Lett. 34: 6337
237. Riva S, Bovara R, Ottolina G, Secundo F, Carrea G (1989) J. Org. Chem. 54: 3161
238. Riva, S, Klibanov AM (1988) J. Am. Chem. Soc. 110: 3291
239. Njar VCO, Caspi E (1987) Tetrahedron Lett. 28: 6549
240. Ottolina G, Carrea G, Riva S (1991) Biocatalysis 5: 131
241. Margolin AL, Delinck DL, Whalon MR (1990) J. Am. Chem. Soc. 112: 2849
242. Delinck DL, Margolin AL (1990) Tetrahedron Lett. 31: 3093
243. Holla EW (1989) Angew. Chem. Int. Ed. Engl. 28: 220
244. Ling L, Ozaki S (1993) Tetrahedron Lett. 34: 2501
245. Henly R, Elie CJJ, Buser HP, Ramos G, Moser HE (1993) Tetrahedron Lett. 34: 2923
246. Lutz D, Guldner A, Thums R, Schreier P (1990) Tetrahedron Lett. 1: 783
247. Morgan B, Oehlschlager AC (1993) Tetrahedron Asymmetry 4: 907
248. Fourneron JD, Chiche M, Pieroni G (!990) Tetrahedron Lett. 31: 4875
249. Geresh S, Elbaz E, Glaser R (1993) Tetrahedron 49: 4939
250. Laane C, Boeren S, Vos K, Veeger C (1987) Biotechnol. Bioeng. 30: 81
251. Zaks A, Klibanov AM (1988) J. Biol. Chem. 263: 3194
252. Parida S, Dordick JS (1991) J. Am. Chem. Soc. 113: 2253
253. Bornscheuer U, Herar A, Kreye L, Wendel V, Capewell A, Meyer HH, Scheper T, Kolisis FN (1993) Tetrahedron Asymmetry 4: 1007
254. Parida S, Dordick JS (1993) J. Org. Chem. 58: 3238
255. Kanerva LT, Vihanto J, Halme MH, Loponen JM, Euranto EK (1990) Acta Chem. Scand. 44: 1032
256. Secundo F, Riva S, Carrea G (1992) Tetrahedron Asymmetry 3: 267
257. Zaks A, Klibanov AM (1988) J. Biol. Chem. 263: 8017
258. Kitaguchi H, Itoh I, Ono M (1990) Chem. Lett. 1203
259. Gutman AL, Shapira M (1991) J. Chem. Soc., Chem. Commun. 1467
260. Adlercreutz P (1991) Eur. J. Biochem. 199: 609
261. Sanchez–Montero JM, Hamon V, Thomas D, Legoy MD (1991) Biochem. Biophys. Acta 1078: 345
262. Kvittingen L, Sjursnes B, Anthonsen T, Halling P (1992) Tetrahedron 48: 2793
263. Yamamoto Y, Kise H (1993) Chem. Lett. 1821
264. Wescott CR, Klibanov AM (1993) J. Am. Chem. Soc. 115: 1629
265. Westcott CR, Klibanov AM (1993) J. Am. Chem. Soc. 115: 10362
266. Terradas F, Teston-Henry M, Fitzpatrick PA, Klibanov AM (1993) J. Am. Chem. Soc. 115: 390
267. Tawaki S, Klibanov AM (1992) J. Am. Chem. Soc. 114: 1882
268. Fitzpatrick PA, Klibanov AM (1991) J. Am. Chem. Soc. 113: 3166
269. Nakamura K, Takebe Y, Kitayama T, Ohno A (1991) Tetrahedron Lett. 32: 4941
270. Ueji S, Fujino R, Okubo N, Miyazawa T, Kurita S, Kitadani M, Muromatsu A (1992) Biotechnol. Lett. 14: 163
271. Rubio E, Fernandez–Mayorales A, Klibanov AM (1991) J. Am. Chem. Soc. 113: 695

Chemical Modification of Proteins with Polyethylene Glycols

Y. Inada, A. Matsushima, M. Hiroto, H. Nishimura and Y. Kodera
Toin Human Science and Technology Center, Department of Materials Science and Technology, Toin University of Yokohama, 1614 Kurogane-cho, Midori-ku, Yokohama, 225 Japan

Proteins as well as bioactive substances were coupled with two types of polyethylene glycol (PEG) derivatives, one with a chain-shaped form, 2,4-*bis*(O-methoxypolyethylene glycol)-6-chloro-*s*-triazine (activated PEG$_2$), and the other with a comb-shaped form, copolymer of PEG and maleic anhydride (activated PM). This modification technique eliminated some of the drawbacks of the native proteins and bioactive substances and also endowed them with new functions. L-Asparaginase and bovine serum albumin caused a reduction of immunoreactivity toward their antibodies and a prolongation of the clearance time through this modification. The modified lipase catalyzed the reverse reaction of hydrolysis, ester synthesis and ester exchange reactions in the hydrophobic media. This technique was further extended to various kinds of bioactive substances to form magnetic urokinase, PEG$_2$-melanin and chlorophyll-bentonite conjugates. These phenomena have been discussed in relation to the structures of the modifiers, PEGs and PMs.

Advances in Biochemical Engineering/
Biotechnology, Vol. 52
Managing Editor: A. Fiechter
© Springer-Verlag Berlin Heidelberg 1995

1 Introduction

Chemical modification of proteins became commonplace in the late 1950s: techniques were developed to facilitate the analysis of the structure-function relationship of the protein molecules. The intention of such modifications was to investigate the physicochemical states of various amino acid residues and to identify the amino acids involved in a particular protein function. Since the 1970s, many articles concerning the chemical modification of proteins by conjugation with synthetic or natural macromolecules have been published with other objectives. The aims of protein modification included the alteration of immunoreactivity or immunogenicity, the suppression of immunoglobulin E production [1], or the production of enzymes soluble in organic solvents [2]. Since polyethylene glycol (PEG), with a general formula of $HO(CH_2CH_2O)_nH$, was synthesized, many industrial and biochemical applications in such fields as pharmaceutics, cosmetics and textiles have been developed to utilize its unique properties [3]. The number of the reports on chemical modification of proteins with PEG has been increasing yearly (Fig. 1).

This review deals with the chemical modification of proteins and enzymes with PEG derivatives to eliminate the drawbacks of the unmodified ones and to improve their properties. This concept was further extended to form various kinds of conjugates, such as PEG-melanin and chlorophylls-montmorillonite.

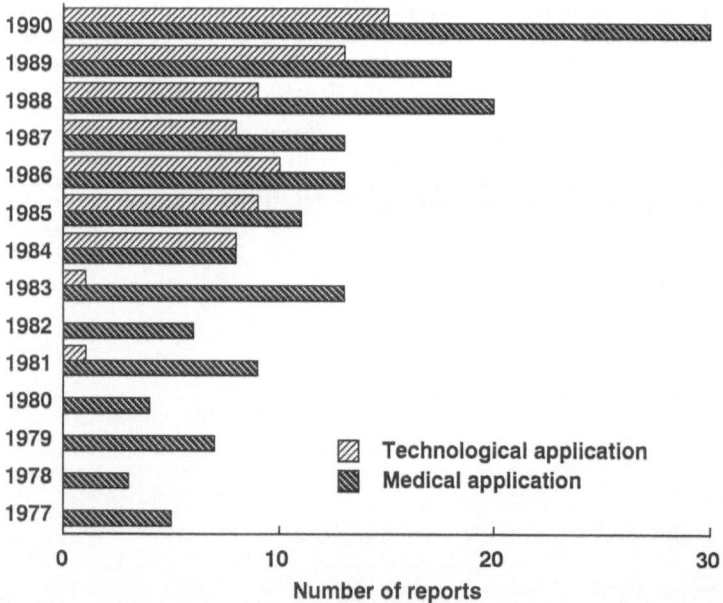

Fig. 1. Number of reports on PEG-conjugates

These conjugates may open a new avenue in medicine, pharmacy, technology and agriculture.

2 Activated Polyethylene Glycols

Proteins can be modified with activated polyethylene glycol derivatives (Fig. 2). The modifier is usually synthesized from monomethoxypolyethylene glycol and cyanuric chloride or N-hydroxysuccinimide [4]. Most of these modifiers have a chain-shaped form, such as 2-(O-methoxypolyethylene glycol)-4,6-dichloro-s-triazine, abbreviated as "activated PEG$_1$" (Fig. 2a); 2,4-bis(O-methoxypolyethylene glycol)-6-chloro-s-triazine, "activated PEG$_2$" (Fig. 2b); and methoxypolyethylene glycol carboxylic acid N-succinimidyl ester, "PEG-succinimide" (Fig. 2c). Recently, we have developed a new type of modifier with a comb-shaped form, a copolymer of PEG and maleic anhydride, abbreviated as "activated PM" (Fig. 2d). Each modifier reacts mainly with the ε-amino group of lysine residue and/or the N-terminal amino group.

Synthesis of activated PEG$_2$. Activated PEG$_2$ (Fig. 2b) was synthesized by Matsushima et al. from methoxypolyethylene glycol (mw: 5000) and cyanuric

<u>chain-shaped PEGs</u>

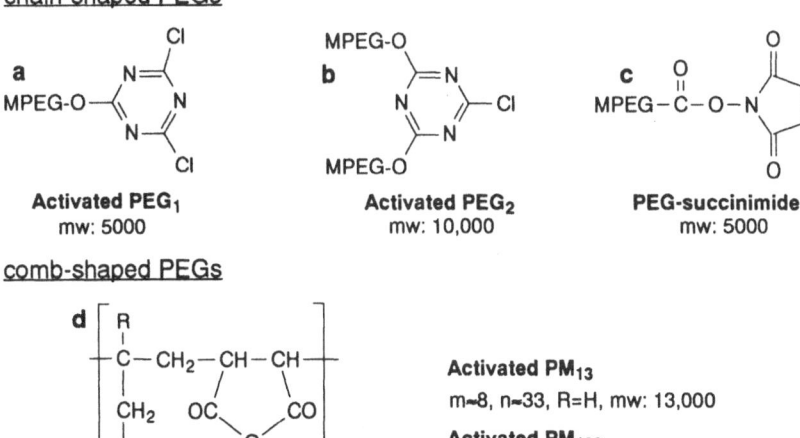

activated PEG$_1$, 2-(O-methoxypolyethylene glycol)-4,6-dichloro-s-triazine
activated PEG$_2$, 2,4-bis(O-methoxypolyethylene glycol)-6-chloro-s-triazine
PEG-succinimide, methoxypolyethylene glycol carboxylic acid succinimidyl ester
activated PM, copolymer of a PEG derivative and maleic anhydride

Fig. 2. Chemical structure of activated PEG derivatives

chloride in benzene using sodium carbonate as the catalyst [5]. Recently, Ono et al. [6] obtained activated PEG_2 in a homogeneous state by using zinc oxide as the catalyst without any byproducts of activated PEG_1 and its polymers (Fig. 3). The modifier, activated PEG_2, has two polyethylene glycol chains per molecule, which can be attached to one amino group. Therefore, a more effective modification of a protein is achieved with activated PEG_2 than with activated PEG_1, which has only one PEG-chain per molecule. Since activated PEG_2 binds to an amino group in a protein through a triazine ring, the binding is rather stable in vitro and in vivo because the modified proteins can easily be degraded by neither chemical nor enzymatic action.

Synthesis of PEG-succinimide. α-Carboxymethyl-ω-methoxypoly(oxyethylene) (mw: 4500) was activated with N-hydroxysuccinimide and dicyclohexylcarbo-diimide to obtain PEG-succinimide (Fig. 2c) [7]. The modifier reacts with the amino groups of a protein at neutral pH.

Synthesis of activated PM. Activated PM, a copolymer of the PEG derivative and maleic anhydride, has a comb-shaped form with multi-valent reactive groups as is shown in Fig. 2d. Two kinds of activated PMs have been prepared by a radical polymerization: activated PM_{13} (mw: 13,000, $m \approx 8$, $n \approx 33$, R = H) and activated PM_{100} (mw: 100 000, $m \approx 50$, $n \approx 40$, R = CH_3). The amino groups in a protein molecule are directly coupled with maleic anhydride in the PM-modifier to form amide bonds.

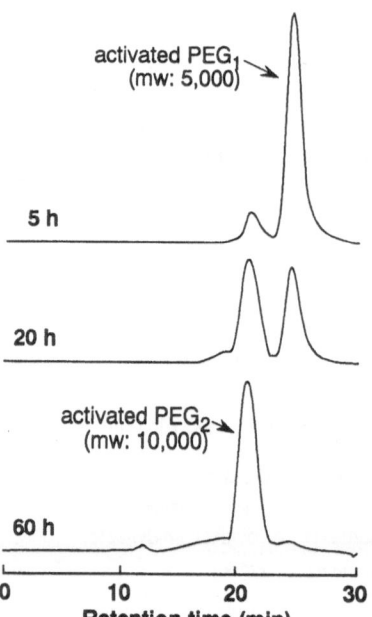

Fig. 3. Time-course of activated PEG_2 synthesis using zinc oxide. The product was analyzed by gel filtration HPLC equipped with TSK gel G3000SW column

3 Medical Application of Protein Hybrids

It is sometimes inevitable to have to administer proteins of non-human origin to humans as protein drugs, although foreign proteins have serious immunogenic side effects such as anaphylactic shock with frequent injections. Many investigators have attempted to reduce the immunogenicity and immunoreactivity of protein drugs without affecting their biological functions. In 1977, Abuchowski et al. [8] demonstrated that catalase modified with activate PEG_1 (Fig. 2a) lost its immunoreactivity towards the anti-serum but retained its enzymic activity. PEG, a non-toxic and non-immunogenic macromolecule, has been used as a protein-precipitation reagent, cell fusion reagent and plasma expander. No serious toxicity was observed when PEG with a molecular weight of 4000 was repeatedly injected into dogs [9]. The modification with PEG has opened new avenues for reducing immunoreactivity and for prolonging the clearance time of foreign protein drugs for biomedical applications. More than two hundred reports of such applications have been published so far. Table 1 lists the recent reports concerning the chemical modification of proteins for biomedical applications.

3.1 Clinical Application of PEG_2-Asparaginase

L-Asparaginase from *Escherichia coli*, which catalyzes the hydrolysis of L-asparagine to L-aspartic acid and ammonia, has been used clinically in leukemia and lymphosarcoma therapy [10].

$$\text{Asparagine} + H_2O \rightarrow \text{Aspartic acid} + NH_3$$

Its administration to patients, however, causes serious immunological side effects. To overcome this drawback, asparaginase was modified by chain-shaped PEG derivatives, activated PEG_1 [11] and activated PEG_2 [5]. These modified asparaginases lost their immunoreactivities and retained their enzymic activities. The anti-tumor activity of asparaginase was markedly enhanced by the modification with activated PEG_2.

The PEG_2-asparaginase was much less susceptible to digestion by trypsin than unmodified asparaginase [5]. Digestion with trypsin for 30 min caused an 80% loss of activity of the unmodified asparaginase, but only a 10% loss of activity of the modified asparaginase. The high resistance of PEG_2-asparaginase to trypsin may be due to the modification of amino groups in lysine residues, which are cleavage sites by trypsin.

Prolongation of clearance time. The clearance time of asparaginase in blood circulation was elongated by the modification of the protein with activated PEG_2. Figure 4 shows the changes in the asparaginase activity and the asparagine level in the sera of rats after administrations of the unmodified and

Table 1. PEG-proteins for biomedical use

---

1 Anti-tumor protein

Antibiotics/Antibody: Yokoyama M et al. (1989) Makromol Chem 190: 2041; Yokoyama M et al. (1990) J Controlled Release 11: 269; Kitamura K et al. (1991) Cancer Res 51: 4310; Takashina K et al. (1991) Jpn J Cancer Res 82: 1145; Yokoyama M et al. (1991) Cancer Res 51: 3229; Nathan A et al. (1993) Bioconjugate Chem 4: 54; Yokoyama M et al. (1993) Pharm Res 10: 895

Asparaginase: Inada Y et al. (1989); In Multiphase Biomedical Materials, VSP, p 153; Cao SG et al. (1990) Ann N Y Acad Sci 613: 460; Teske E et al. (1990) Eur J Cancer 26: 891; Wada H et al. (1990) Ann N Y Acad Sci 613: 95; Kodera Y et al. (1992) Biochem Biophys Res Commun 184: 144

Colony-stimulating factor: Knüsli C et al. (1990) Exp Hematol 18: 608; Tanaka H et al. (1991) Cancer Res 51: 3710; Knüsli C et al. (1992) Br J Haematol 82: 654; Malik F et al. (1992) Exp Hematol 20: 1028

γ-Interferon: Kita Y et al. (1990) Drug Des Delivery 6: 157

Interleukin 2: Zimmerman RJ et al. (1989) Cancer Res 49: 6521; Goodson RJ et al. (1990) Bio/Technology 8: 343; Katre NV (1990) J Immunol 144: 209; Kunitani M et al. (1991) J Chromatogr 588: 125; Teppler H et al. (1993) J Exp Med 177: 483; Teppler H et al. (1993) J Infect Dis 167: 291

2 Inherited deficiency of enzyme

Bilirubin oxidase: Sugi K et al. (1989) Biochim Biophys Acta 991: 405; Kimura M et al. (1990) Proc Soc Exp Biol Med 195: 64

Gulonolactone oxidase: Hadley KB et al. (1989) Enzyme 42: 225

Purine nucleoside phosphorylase: Hershfield MS et al. (1991) Proc Natl Acad Sci USA 88: 7185

Uricase: Yasuda Y et al. (1990) Chem Pharm Bull 38: 2053

3 Anti-inflammation

Catalase: Liu TH et al. (1989) Am J Physiol 256: H589; White CW et al. (1989) J Appl Physiol 66: 584; Greenwald RA (1990) Free Rad Biol Med 8: 201; Jacobson JM et al. (1990) J Appl Physiol 68: 1252; Katsumata U et al. (1990) Am Rev Respir Dis 141: 1158; Mossman BT et al. (1990) Am Rev Respir Dis 141: 1266; Walther FJ et al. (1990) Exp Lung Res 16: 177; Greenwald RA (1991) Free Rad Res Comms 12–13: 531; Thibeault DW et al. (1993) Exp Lung Res 19: 137; Walther FJ et al. (1993) Pediatr Res 33: 332

Superoxide dismutase: Chi L et al. (1989) Circ Res 64: 665; Veronese FM et al. (1989) J Controlled Release 10: 145; White CW et al. (1989) J Appl Physiol 66: 584; Banci L et al. (1990) J Inorg Biochem 39: 149; Cao SG et al. (1990) Ann N Y Acad Sci 613: 460; Tanaka M et al. (1990) Circ Res 67: 636; Veronese FM et al. (1990) Ann N Y Acad Sci 613: 468; Walther FJ et al. (1990) Exp Lung Res 16: 177; Zhang XM et al. (1990) Am J Physiol 259: H497; Greenwald RA (1991) Free Rad Res Comms 12–13: 531; Haun SE et al. (1991) Stroke 22: 655; Omar BA et al. (1991) J Mol Cell Cardiol 23: 149; Somack R et al. (1991) Free Rad Res Comms 12–13: 553; Hiraoka Y et al. (1992) Cardiovasc Res 26: 956; Kawasaki S et al. (1992) Jpn J Cancer Res 83: 899; Snider J et al. (1992) J Chromatogr 599: 141; Stone WC et al. (1992) Am J Vet Res 53: 2153; Cosenza C et al. (1993) Transplant Proc 25: 1881; Tang G et al. (1993) J Appl Physiol 74: 1425; Thibeault DW et al. (1993) Exp Lung Res 19: 137; Walther FJ et al. (1993) Pediatr Res 33: 332

4 Suppression of immune response

Ovalbumin: Chen Y et al. (1992) Cell Immunol 142: 16

Review: Sehon AH (1991) in Immunobiology of Proteins and Peptides 4, Plenum, p 199

5 Anti-thrombosis

Hirudin: Kurfürst MM (1992) Anal Biochem 200: 244; Zawilska K et al. (1993) Thromb Res 69: 315

Streptokinase: Brucato FH et al. (1990) Blood 76: 73

Thrombin: Nakagomi K et al. (1990) Biochem Int 22: 75

Trypsin (Serine protease): Takoi K et al. (1989) Agric Biol Chem 53: 2063; Zalipsky S et al. (1992) Biotechnol Appl Biochem 15: 100

Urokinase: Yoshimoto T et al. (1989) Drug Delivery System 4: 121

6 Blood substitute

Albumin: Head DM et al. (1989) Biotechnol Tech 3: 27; Delgado C et al. (1990) Biotechnol Appl Biochem 12: 119; Yoshinaga K et al. (1990) J Appl Polym Sci 41: 1443; Bergström K et al. (1991) Biotechnol Bioeng 38: 952; Matsushima A et al. (1992) Biochem Int 26: 485; Sasaki H et al. (1992) Biochem Biophys Res Commun 197: 287

Table 1. (Continued)

Hemoglobin: Malchesky PS et al. (1990) Int J Artif Organs 13: 442; Yabuki A et al. (1990) Transfusion 30: 516; Iwashita Y (1991) Artif Organs Today 1: 89; Kida Y et al. (1991) Artif Organs 15: 5; Gotoh K et al. (1992) Artif Organs 16: 586; Matsumura H et al. (1992) Artif Organs 16: 461; Ohno H (1992) Electrochim Acta 37: 1649

Immunoglobulin: Kondo A et al. (1989) Biotechnol Bioeng 34: 532; Suzuki T et al. (1989) J Biomater Sci Polym Ed 1: 71; Kunitani M et al. (1991) J Chromatogr 588: 125; Sada E et al. (1991) J Ferment Bioeng 71: 137; Cunningham-Rundles C et al. (1992) J Immunol Methods 152: 177

Myoglobin: Kunitani M et al. (1991) J Chromatogr 588: 125; Ohno H et al. (1992) J Electroanal Chem 341: 137

Protease inhibitor: Mast AE et al. (1990) Biol Chem Hoppe-Seyler 371(Suppl.): 101; Mast AE et al. (1990) J Lab Clin Med 116: 58

7 Miscellaneous

Alkaline phosphatase: Harris JM et al. (1989) J Bioact Compat Polym 4: 281; Yoshinaga K et al. (1989) J Bioact Compat Polym 4: 17

Fetuin: Roseng L et al. (1992) J Biol Chem 267: 22987

Lipid: Papahadjopoulos D et al. (1991) Proc Natl Acad Sci USA 88: 11460; Maruyama K et al. (1992) Biochim Biophys Acta 1128: 44; Woodle MC et al. (1992) Int J Pharm 88: 327; Gabizon AA et al. (1993) Pharm Res 10: 703; Silvius JR et al. (1993) Biochemistry 32: 3153; Zalipsky S (1993) Bioconjugate Chem 4: 296

Nucleotide: Jäschke A et al. (1993) Tetrahedron Lett 34: 301

Peptide: Head DM et al. (1989) Biotechnol Tech 3: 27; Atassi MZ et al. (1991) J Protein Chem 10: 623; Goddard P et al. (1991) J Pharm Sci 80: 1171; Kawasaki K et al. (1991) Chem Pharm Bull 39: 3373; Powers SP et al. (1991) Biochemistry 30: 676; Braatz JA et al. (1993) Bioconjugate Chem 4: 262

Protease: Narukawa H et al. (1993) FEMS Microbiol Lett 108: 43

Protein A: Head DM et al. (1989) Biotechnol Tech 3: 27

Ribonuclease: Caliceti P et al. (1990) J Mol Recognit 3: 89; Kunitani M et al. (1991) J Chromatogr 588: 125; Schiavon O et al. (1991) Farmaco 46: 967

Soybean trypsin inhibitor: Takakura Y et al. (1989) J Pharm Sci 78: 219; Gaertner HF et al. (1990) Biotechnol Bioeng 36: 601

Aqueous two-phase system: Lei X et al. (1992) J Chromatogr 15: 2801

8 Review

Zalipsky S et al. (1991) ACS Symp Ser 469: 91; Zalipsky S et al. (1991) Polym Prepr 31: 173; Fuertges F et al. (1990) J Controlled Release 11: 139; Nucci ML et al. (1991) Adv Drug Delivery Rev 6: 133

PEG_2-asparaginases [54]. After the intraperitoneal injection of PEG_2-asparaginase, its activity appeared at its peak level for 10–20 h. The activity decreased far more slower than that of unmodified asparaginase and was still detectable after three weeks. Correspondingly, the asparagine level in the serum was still undetectable three weeks later, although that treated with the unmodified enzyme increased after only 20 h. This depletion of asparagine persisted for more than 20-times longer than that after an injection of unmodified asparaginase. The half-lives of the unmodified and modified asparaginases in the serum were calculated to be 2.9 and 56 h, respectively. The increased circulatory lifetime of the modified asparaginase may be due to its resistance to the reticuloendothelial system, which is responsible for the elimination of foreign proteins.

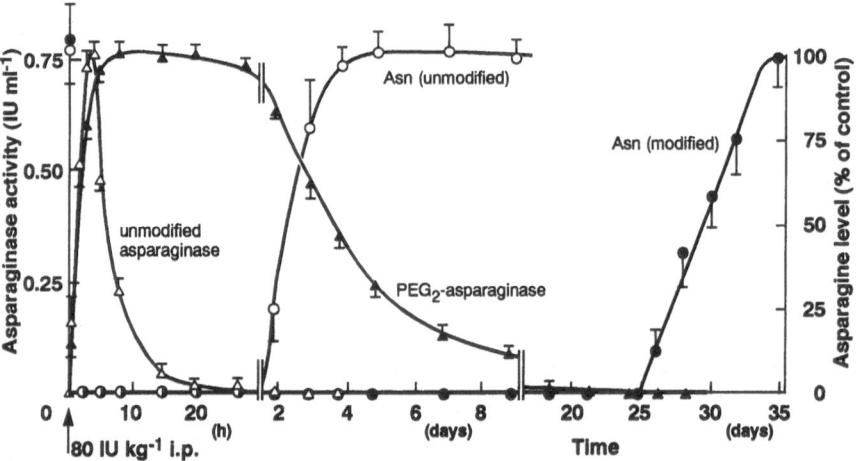

Fig. 4. Changes in L-asparaginase activity and L-asparagine concentration in rat sera after intraperitoneal injection of unmodified- and PEG_2-asparaginases

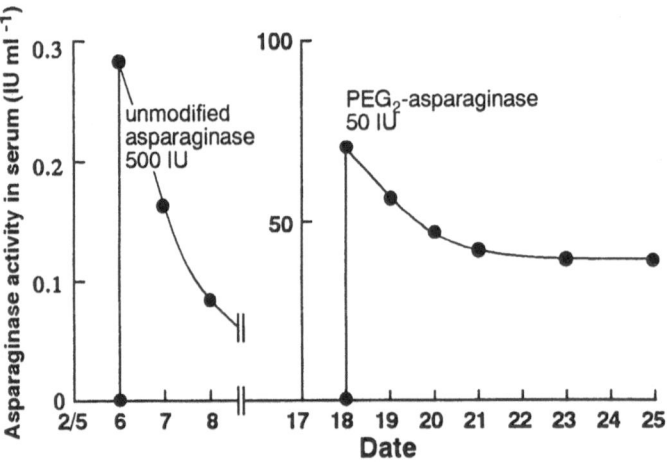

Fig. 5. Clearance times of L-asparaginase and PEG_2-asparaginase and their therapeutic effectiveness in a lymphosarcoma-bearing dog. The dog was treated intravenously with unmodified asparaginase (500 IU) and then with modified asparaginase (50 IU; see Table 2)

Preclinical trial. An increased anti-tumor activity of the modified asparaginase in vivo could be expected from the half-life of this enzyme and from the persistence of high enzymic activity in the serum after its injection. Figure 5 and Table 2 show a therapeutic study of unmodified and PEG_2-asparaginases in a dog with a spontaneous lymphosarcoma [12]. Asparaginase activity appeared in the serum after injection of unmodified asparaginase and decreased with a

Table 2. Therapeutic effectiveness of PEG_2-asparaginase in a lymphosarcoma-bearing dog[a]

Date: February		Unmodified asparaginase (500 IU) iv				PEG_2-asparaginase (50 IU) iv				
		5	6	7	8	17	18	19	23	25
Serum level (µM)	Asp		21	92			10	35	59	69
	Asn		75	nd			73	nd	nd	nd
	Glu		158	336			82	181	203	346
	Gln		796	1211			574	633	866	977
No. in peripheral	$RBC \times 10^4$	283				280				345
blood (per mm³)	$WBC \times 10^3$	26				20				11
Ascites		+ + +			+	+ + +		−	−	−
Lymph nodes	(cm × cm)	4 × 4			2 × 2	3 × 4		2 × 3	0 × 0	0 × 0

[a] The dog was treated intravenously with unmodified asparaginase (500 IU) and then PEG_2-asparaginase (50 IU). On the date indicated, blood was taken from a peripheral vein to measure the serum amino acid levels and a count of the red blood cells (RBC) and white blood cells (WBC) was made. The values of some amino acids in the serum shown on 6 and 18 February were obtained before administration of the unmodified and modified enzyme, respectively. Abbreviation: nd, not detected (less than 0.5 µM)

half-life of roughly 1 day. During this time, the size of the neck lymph nodes and the volume of ascites were reduced, indicating that the tumor·was asparaginase-sensitive. During the 12 days after the first injection, the lymph nodes again increased in size due to the complete disappearance of the enzymic activity and the elevation of the asparagine level in the serum. However, an injection of PEG_2-asparaginase at one-tenth of the enzymic activity of that in the first injection caused striking shrinkage of the neck lymphosarcoma and a reduction of the ascites due to the complete depletion of serum asparaginase. The lymph nodes became nonpalpable within 5 days after the second injection. The clearance time of PEG_2-enzyme in the blood was much longer than that of the unmodified enzyme.

PEG_2-asparaginase was given to four patients with leukemia who had developed an allergic reaction to asparaginase during the methotrexate-asparaginase sequential maintenance therapy [13]. They have received the regimen without any allergic reaction or side effects such as liver dysfunction, coagulation disorder, or diabetes mellitus, which is attributable to asparaginase.

3.2 Modification of L-Asparaginase with Activated PM

Recently a comb-shaped copolymer of PEG and maleic anhydride, activated PM was developed as a modifier (Fig. 2d) [14]. Figure 6 presents the degree of modification, the immunoreactivity and the enzymic activity of asparaginases modified with activated PM_{13} (Fig. 6a) and activated PM_{100} (Fig. 6b) [15]. In both cases, the immunoreactivity sharply decreased with increasing the degree of modification and approached zero.

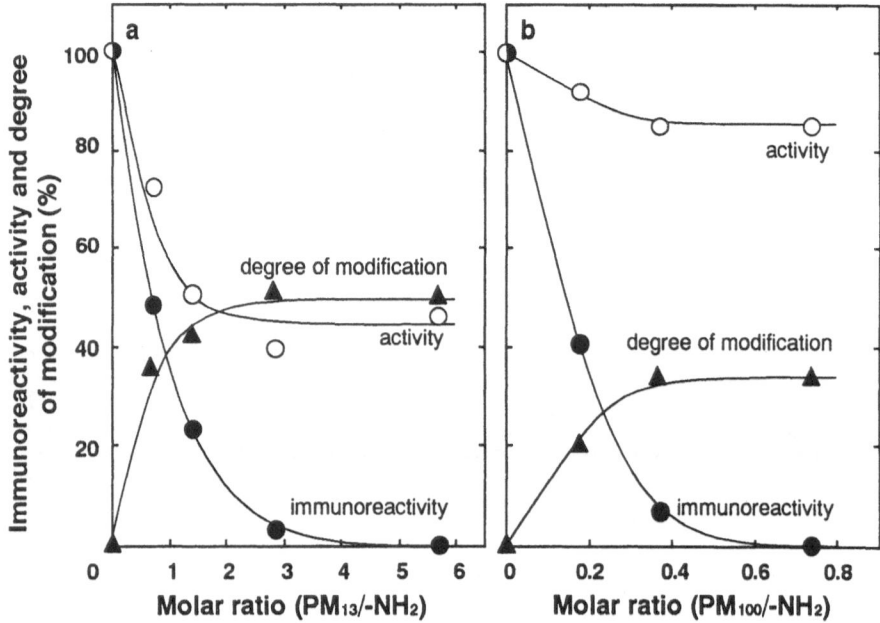

Fig. 6a, b. Degree of modification, immunoreactivity and enzymic activity of modified asparaginases: **a** PM_{13}-asparaginase; **b** PM_{100}-asparaginase

Chain-shaped and Comb-shaped PEGs. Table 3 shows the degree of modification and the enzymic activity of the modified asparaginases when their immunoreactivities were completely lost on coupling with various kinds of modifying reagents. The two comb-shaped copolymers, activated PM_{13} and PM_{100}, were more effective than the chain-shaped polymers, activated PEG_1 [11] and PEG_2 [5]; the residual activities of PM_{13}- and PM_{100}-asparaginases were 45% and 85% of the unmodified enzyme, respectively. Comparing activated PEG_2 (mw: 10 000) with activated PM_{13} (mw: 13 000) as modifying reagents, PM_{13}-asparaginase without immunoreactivity exhibited a higher enzymic activity (45%) with a lower degree of modification (50%) than that (11% activity with 57% modification) of PEG_2-asparaginase without immunoreactivity. The comb-shaped polymers with many reactive groups of maleic anhydride react directly with the ε-amino groups in lysine residues and N-terminal amino groups in a protein molecule. Furthermore, they may cover the whole surface of the protein molecule with hydrogen bonds between side chains of amino acid residues and oxygen atoms in the branching PEG chains.

Asparaginase modified with activated PM_{100} showed a higher enzymic activity and a lower degree of modification of amino groups accompanied by the complete loss of immunoreactivity, than those of other modifiers such as activated PEG_1, PEG_2 and PM_{13} (Table 3). The presumptive molecular sizes of the activated PMs are shown in Table 4. Activated PM_{100} has a main chain 260-

Table 3. Activity and immunoreactivity of modified asparaginases

Modifying reagent (molecular weight)	Degree of modification (%)	Enzymic activity (%)	Immuno-reactivity (%)
unmodified asparaginase	0 (0)[a]	100[b]	100
activated PEG$_1$ (5 000)	79 (73)	0.9	0
activated PEG$_2$ (10 000)	57 (52)	11.0	0
activated PM$_{13}$ (13 000)	50 (46)	45	0
activated PM$_{100}$ (100 000)	34 (31)	85	0

[a] Parentheses: the number of amino groups coupled with each modifying reagent. The total number of amino groups in the asparaginase molecule is 92
[b] The activity of the unmodified enzyme is 200 U mg^{-1} protein by the GOT method

Table 4. Comparison of activated PMs in their molecular sizes

	Presumptive molecular size[a]	
	Main chain (**m**)	Branching PEG chain (**n**)
activated PM$_{13}$	40 Å (8)	64 Å (33)
activated PM$_{100}$	260 Å (50)	78 Å (40)

[a] The lengths of the chains were calculated assuming the PEG chain to be a meandering form [3]. In parentheses: degree of polymerization

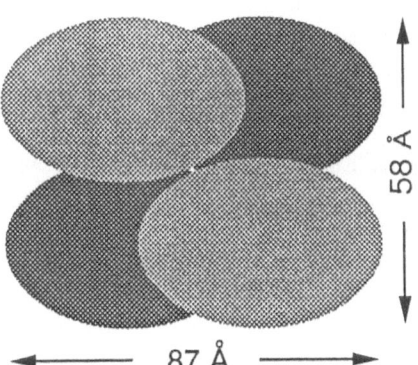

Fig. 7. Conformation of L-asparaginase from *E. coli*

Å long, which is more than six times longer than that of activated PM$_{13}$. Figure 7 shows the rough conformation of L-asparaginase from *E. coli* as measured by X-ray diffraction [16]. Its molecular weight is 136 000 and the molecule consists of four identical subunits. The length of the main chain of activated PM$_{100}$, 260 Å, may be enough to cover the whole surface of the globular asparaginase molecule (87 × 63 × 58 Å), ranging from 180–270 Å in circumference.

Table 5. Activity and immunoreactivity of modified BSAs

Modifying reagent (molecular weight)	Degree of modification (%)	Enzymic activity (%)	Immuno- reactivity (%)
unmodified BSA	0 (0)[a]	100[b]	100
activated PEG_1[c] (5 000)	42 (25)	—	0
activated PEG_2 (10 000)	25 (15)	63	0
activated PM_{13} (13 000)	30 (18)	63	0
activated PM_{100} (100 000)	20 (12)	93	0

[a] Parentheses: the number of amino groups coupled with each modifying reagent. The total number of amino groups in BSA is 60
[b] The activity of the unmodified BSA is 30 $\mu mol\, min^{-1}\, g^{-1}$ using p-nitrophenyl acetate as the substrate
[c] The esterase activity was not determined

3.3 Modification of Bovine Serum Albumin with PEG Derivatives

To confirm the predominance of activated PM_{100} in protein modification, bovine serum albumin, BSA, was coupled with various types of PEG derivatives, using BSA as a model protein. The reduction of immunoreactivity of BSA modified with various kinds of modifiers was studied in relation to the degree of modification and the esterase activity of the conjugates [55]. It was reported that BSA has an esterase activity and catalyzes the hydrolysis of p-nitrophenyl acetate [17].

Table 5 shows the esterase activity of the modified BSA preparations which had completely lost its immunoreactivity towards the antiserum. Abuchowski et al. [18] reported that modified BSA, in which 25 of 60 amino groups (42%) were coupled with activated PEG_1, did not form a precipitin line on the immunodiffusion plate. Using activated PEG_2 and activated PM_{13}, the enzymic activities retained were 63% with the modification of 15 and 18 amino groups, respectively. In the case of PM_{100}-BSA, the activity was almost completely retained by 93% of the original activity with 20% modification of amino groups.

BSA is a prolate ellipsoid with major and minor axes of 141 and 41 Å and consists of a single peptide chain with a molecular weight of 66 267. Therefore activated PM_{100} may cover the whole surface of the BSA molecule with a lower degree of modification.

4 Biotechnological Application of Protein Hybrids

The use of enzymes in organic solvents greatly expands their potential applications in organic synthesis [19]. To accomplish this, many investigators have developed a lot of techniques; these include the suspension of enzymes in organic

Table 6. PEG-enzymes for biotechnological use

---

1 Oxidoreductase
Catalase: Miyamoto M et al. (1990) Macromolecules 23: 3201
Dehydrogenases: Yomo T et al. (1991) Ann N Y Acad Sci 613: 313; Yomo T et al. (1991) Eur J
    Biochem 196: 343
Glucose oxidase: Yabuki S et al. (1992) Biosens Bioelectron 7: 695
Peroxidase: Souppe J et al. (1989) New J Chem 13: 503

2a Esterase
Lipase: Kikkawa S et al. (1989) Biochem Int 19: 1125; Mizutani A et al. (1989) J Biotechnol 10: 121;
    Stark MB et al. (1989) Biotechnol Bioeng 34: 942; Inada Y et al. (1990) Biocatalysis 3: 317; Basri M
    et al. (1991) Biocatalysis 4: 313; Hiroto M et al. (1992) Biotechnol Lett 14: 559; Kodera Y et al.
    (1993) J Biotechnol 31: 219
Phospholipase: Matsuyama H et al. (1991) Chem Pharm Bull 39: 743; Matsuyama H et al. (1992)
    Chem Pharm Bull 40: 2478

2b Protease
Alkaline proteinase: Yamagata Y et al. (1989) Curr Microbiol 19: 307
Chymotrypsin: Babonneau MT et al. (1989) Tetrahedron Lett 30: 2787; Gaertner H et al. (1989) Eur J
    Biochem 181: 207; Pina C et al. (1989) Biotechnol Tech 3: 333; Fulcrand V et al. (1990) Tetrahedron
    46: 3909; Nakajima A et al. (1990) J Biomater Sci Polym Ed 1: 183; Sakurai K et al. (1990)
    Biotechnol Lett 12: 685; Fulcrand V et al. (1991) Int J Peptide Protein Res 38: 273; Gaertner H et al.
    (1991) J Org Chem 56: 3149; Cabezas MJ et al. (1992) J Mol Catal 71: 261; Sanchez-Montero JM et
    al. (1992) Prog Biotechnol 8: 371; Chiu HC et al. (1993) Bioconjugate Chem 4: 290
Papain: Lee HH et al. (1989) Chem Express 4: 253; Ohwada K et al. (1989) Biotechnol Lett 11: 499;
    Souppe J et al. (1989) New J Chem 13: 503; Gaertner HF et al. (1990) Biocatalysis 3: 197; Lee HH
    (1990) Chem Express 5: 469; Nakajima A et al. (1990) J Biomater Sci Polym Ed 1: 183; Sakurai K et
    al. (1990) Biotechnol Lett 12: 685; Uemura T et al. (1990) Agric Biol Chem 54: 2277; Jayakumari
    VG et al. (1991) Proc Indian Acad Sci Chem Sci 103: 133; Sanchez-Montero JM et al. (1992) Prog
    Biotechnol 8: 371
Pepsin: Sakurai K et al. (1990) Biotechnol Lett 12: 685
Subtilisin: Gaertner H et al. (1989) Eur J Biochem 181: 207; Ferjancic A et al. (1990) Appl Microbiol
    Biotechnol 32: 651
Thermolysin: Sakurai K et al. (1990) Biotechnol Lett 12: 685; Sanchez-Montero JM et al. (1992) Prog
    Biotechnol 8: 371
Trypsin: Gaertner H et al. (1989) Eur J Biochem 181: 207; Gaertner HF et al. (1990) Biotechnol
    Bioeng 36: 601; Nakajima A et al. (1990) J Biomater Sci Polym Ed 1: 183; Sakurai K et al. (1990)
    Biotechnol Lett 12: 685; Munch O et al. (1991) Biocatalysis 5: 35; Caliceti P et al. (1993) J Bioact
    Compat Polym 8: 41

2c Glycosidase
Cellulase: Garcia A et al. (1989) Biotechnol Bioeng 33: 321
β-*Galactosidase*: Beecher JE et al. (1990) Enzyme Microb Technol 12: 955

3 Miscellaneous
NAD(P): Yomo T et al. (1989) Eur J Biochem 179: 299; Hanson RL et al. (1990) Bioorg Chem 18: 116;
    Ottolina G et al. (1990) Enzyme Microb Technol 12: 596; Scheper T et al. (1990) Biosens
    Bioelectron 5; 125; Guagliardi A et al. (1991) Biotechnol Appl Biochem 13: 25; Kulys JJ et al. (1991)
    Anal Lett 24: 181; Yomo T et al. (1991) Ann N Y Acad Sci 613: 313; Yomo T et al. (1991) Eur J
    Biochem 196: 343
Lignin: Cole BJW et al. (1993) J Wood Chem Technol 13: 59

---

media [20] or formation of lipid-enzyme complex [21]. These techniques have
some advantages and disadvantages for the biotechnological application of
enzymes. During the course of these investigations on enzyme-catalyzed reaction in organic solvents, Inada et al. found that enzymes modified with PEG
become soluble and retain the enzymic activity in organic solvents [1, 22–24].

Since PEG is an amphipathic macromolecule, its hydrophilic nature makes it possible to modify enzymes in aqueous solution, and its hydrophobic nature would make modified enzymes soluble in a hydrophobic environment. In fact, modified enzymes such as catalase [25] and peroxidase [26] become soluble in benzene and in chlorinated hydrocarbons, and exhibit markedly high activities in these organic solvents. Furthermore, the PEG-modified hydrolytic enzymes such as lipase [27] effectively catalyze the reverse reaction of hydrolysis in hydrophobic media – ester synthesis and ester exchange reactions in organic solvents:

$$R_1COOH + R_2OH \rightarrow R_1COOR_2 + H_2O$$

$$R_1COOR_2 + R_3COOR_4 \rightarrow R_1COOR_4 + R_3COOR_2.$$

In the case of proteases such as PEG_2-papain [28] and PEG_2-chymotrypsin [29], the amide bond formation is also catalyzed in organic solvents:

$$R_1COOH + R_2NH_2 \rightarrow R_1CONHR_2 + H_2O.$$

Table 6 shows the recent reports on PEG-enzymes for biotechnological applications. The success in the enzyme modification with PEG derivatives depends upon the degree of modification of the amino groups in each enzyme molecule. Too low a degree of modification would not bring about sufficient solubilization of proteins in organic solvents, while too high a degree of modification would cause needless reduction of the enzymic activity because of the distortion of the protein conformation. The degree of modification of the amino groups in a protein molecule can be adjusted by changing the amount of the modifier added to the enzyme solution.

4.1 Properties of PEG-Lipases

Preparation of PM_{13}-lipase. Lipase from *Pseudomonas fluorescens* was coupled with activated PM_{13} (Fig. 2d) to obtain PM_{13}-lipase [30]. The degree of modification was increased by increasing the molar ratio and tended to approach a constant level of 60% (Fig. 8). The activity in an emulsified aqueous system of PM_{13}-lipase was 1.3 times higher than that of unmodified lipase.

PM_{13}-lipase becomes soluble in organic solvents, so that the reverse reactions of hydrolysis, ester synthesis reactions, are catalyzed in the solvents. In fact, lauryl stearate was synthesized in a clear benzene solution containing lauryl alcohol and stearic acid with PM-lipase. While the unmodified lipase did not catalyze any ester synthesis reaction, the ester synthesis activity of PM-lipase was enhanced by increasing the degree of modification and tended to approach a constant level ($4.7 \mu mol\, min^{-1}\, mg^{-1}$). So far, lipase from *P. fluorescens*, carboxyl groups formed by the hydrolysis of activated PM do not serve as a substrate of ester synthetic reaction in organic solvents, which is consistent with its substrate specificity [27].

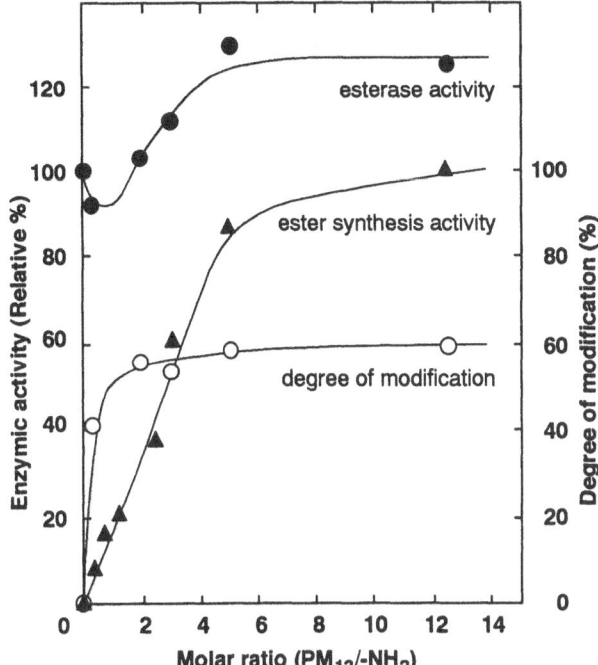

Fig. 8. Degree of modification of PM_{13}-lipase together with esterase activity in an emulsified system and ester synthesis activity in benzene. The modification was performed at 4 °C for 1 h

Heat-stability of PM_{13}-lipase. PM_{13}-lipase [molar ratio $(PM_{13}/-NH_2) = 12.5$ in Fig. 8] as well as the unmodified lipase was incubated at 55 °C [30]. Although the esterase activity of unmodified lipase in an emulsified system was completely lost in a 150-min incubation, 60% of the activity of PM_{13}-lipase was retained even after 150 min. Similarly, 50% of the ester synthesis activity of PM_{13}-lipase in benzene was retained in the same incubation time.

Properties of PEG_2-lipase. Lipase from *P. fluorescens* modified with activated PEG_2 (Fig. 2b) exhibited the enzymic activity in water-immiscible organic solvents such as benzene, toluene, chloroform and 1,1,1-trichloroethane. The ester synthetic activity of PEG_2-lipase in 1,1,1-trichloroethane was almost 3.6 times higher than that in benzene or toluene [31]. However, a trace amount of water in the water-immiscible organic solvents was needed to exhibit these activities. Little activity was detected in water-miscible solvents such as acetone and dimethylformamide [32]. The water molecules may be absorbed at the surface of the protein molecule so that the protein conformation could be maintained. Therefore the organic solvents saturated with water are recommended in reverse reactions of hydrolysis with PEG_2-lipase.

PEG_2-lipase exhibits a high stability in benzene; PEG_2-lipase still retained about 50% of its original activity for amyl laurate synthesis after a 3-month storage at room temperature, and about 40% activity even after 140 days [33]. PEG_2-lipase can be recovered from the reaction mixture as a precipitate by adding *n*-hexane or petroleum ether [33].

4.2 Ester Synthesis and Ester Exchange Reactions with PEG_2-Lipase in Organic Solvents

Ester exchange reaction without oxidation of substrates. Ester exchange reactions between retinyl acetate and palmitic acid or oleic acid were performed with PEG_2-lipase in benzene under nitrogen gas at $25\,°C$ [34]. For comparison, retinyl ester syntheses were performed by conventional organic reaction using p-toluenesulfonic acid under nitrogen gas for a 3-h refluxing. As much as 85% of retinyl acetate was converted to retinyl palmitate with PEG_2-lipase, and its peroxide value (2.5 meq kg^{-1}) was almost 17 times lower than the 43 meq kg^{-1} obtained by organic synthesis (about 30% yield). A similar result was obtained for retinyl oleate synthesis in benzene: 9.0 meq kg^{-1} for enzymatic reaction and 200 meq kg^{-1} for organic synthesis. Obviously ester exchange reaction from two substrates with double bonds using modified-enzyme is more preferable than that by the organic synthesis.

PEG_2-lipase in neat substrate. Since PEG_2-lipase is soluble not only in organic solvents but also in hydrophobic substrates, ester synthesis and ester exchange reactions are catalyzed with PEG_2-lipase in hydrophobic substrates without organic solvents [35]. In the mixture of the two substrates of $1.35\,M$ amyl laurate and $2.7\,M$ lauryl alcohol in the presence of PEG_2-lipase, lauryl laurate was synthesized with a specific activity of 2.6 $\mu mol\,min^{-1}\,mg^{-1}$. The optimum temperature of this reaction was $65\,°C$, although that of hydrolysis-reaction in an aqueous emulsified state with unmodified lipase is $45\,°C$. The higher optimum temperature shown by PEG_2-lipase may be due to the modification with activated PEG_2 and/or to the alteration of the reaction environment. PEG_2-lipase efficiently catalyzed ester exchange reactions between two kinds of triglycerides in neat liquid substrates [36]. The versatile activity and heat stable nature, as well as the solubility in neat hydrophobic substrate, make the PEG_2-lipase extremely useful for many practical applications such as the alteration of the properties of fats and oils.

4.3 Kinetic Studies of PEG-Lipases in Organic Solvents

The kinetic studies of the reverse reactions of hydrolysis in organic solvents were tested with the modified lipases. The enzymic activity of an ester synthesis reaction is increased with increase of the substrate concentration and tends to reach a constant level. The reciprocal plot of the substrate concentration (S^{-1}) against the reaction rate (v^{-1}) gives a straight line, from which apparent V_{max} and K_m values together with K_i values of inhibitors can be obtained.

Kinetic studies of modified lipases. Table 7 shows the K_m, K_i and V_{max} values of PEG_2-lipase from *P. fluorescens* for various alcohols and fatty acids [27]. The V_{max} value was enhanced by increasing the carbon number of either fatty acid or alcohol, while the K_m value was little affected. Fatty acids with a branching

Table 7. K_m and V_{max} values of ester synthesis by PEG$_2$-lipase from *P. fluorescens* together with K_i value of inhibitors in benzene at 20 °C

Substrates	K_m (M)	K_i (M)	V_{max} (μmol min^{-1} mg^{-1})
Pentyl alcohol (0.75 M)			
pentanoic acid	0.17		2.8
hexanoic acid	0.22		8.0
octanoic acid	0.14		8.1
dodecanoic acid	0.31		10.3
2-methylpentanoic acid[a]	—	0.22	—
3-methylpentanoic acid[a]	—	0.27	—
4-methylpentanoic acid	0.19		0.2
benzoic acid[a]	—	0.10	—
3-phenylpropionic acid	0.18		7.0
4-phenylbutyric acid	0.15		7.3
Pentanoic acid (0.50 M)			
pentyl alcohol	0.23		2.6
hexyl alcohol	0.25		2.2
octyl alcohol	0.24		5.4
dodecyl alcohol	0.33		8.0
2-methylpentyl alcohol	0.21		2.5
3-methylpentyl alcohol	0.62		2.0
4-methylpentyl alcohol	0.19		1.0
1-methylbutyl alcohol	0.31		0.1
1,1-dimethylpropyl alcohol[b]	—	—	—

[a] Acted as inhibitors, not as substrates
[b] Acted neither as an inhibitor nor as a substrate

carbon chain at the position neighboring the carboxyl group acted as competitive inhibitors of ester synthesis from pentyl alcohol and pentanoic acid. With regard to another substrate, alcohols, the substrate specificity was not so restrictive as with fatty acids but secondary and tertiary alcohols did not serve as a substrate.

However, the results shown here were obtained from *P. fluorescens*, and the kinetic properties may be different in the kind of lipases depending on the different sources. Figure 9 presents the substrate specificities of lipase from *P. fragi* modified with activated PEG$_2$ [37] and that from *Candida cylindracea* modified with PEG-succinimide (Fig. 2c) [38] together with PEG$_2$-lipase from *P. fluorescens*. The substrate specificity of *C. cylindracea* lipase for *n*-alcohols and fatty acids was quite different from that of PEG$_2$-lipases from *Pseudomonas* species.

Optical resolution. From a kinetic study of the esterification of chiral secondary alcohols, PEG$_2$-lipase from *P. fragi* preferentially catalyzed the acylation of (*R*)-isomers [39]. The optical resolution of (*RS*)-α-phenylethanol was performed with *n*-dodecanoic acid and PEG$_2$-lipase in 1,1,1-trichloroethane, and only (*R*)-isomer was esterified with PEG$_2$-lipase. The optical purity of the unreacted alcohol, (*S*)-α-phenylethanol, reached a 99% enantiomeric excess in 7 h.

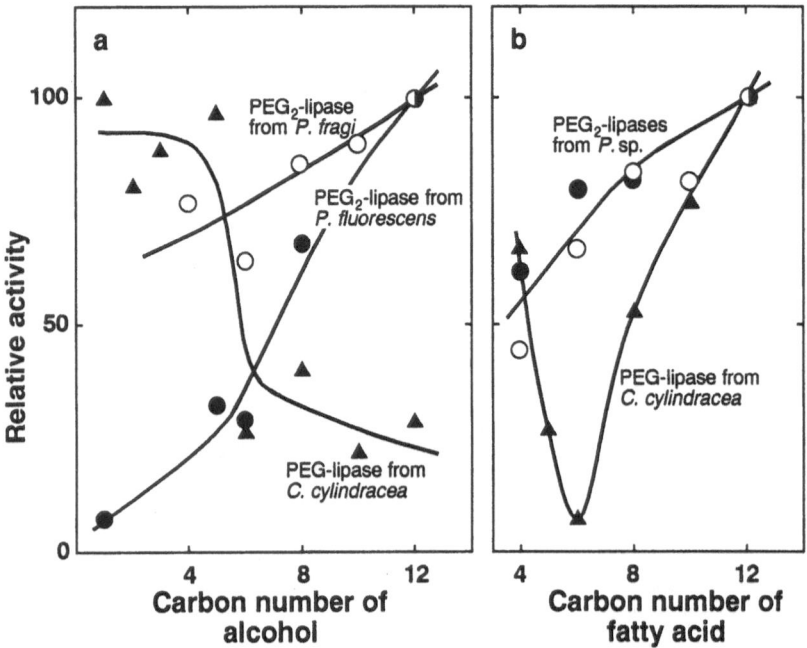

Fig. 9a, b. Substrate specificities of modified lipases from *P. fluorescens*, *P. fragi* and *C. cylindracea*: **a** varying carbon numbers of alcohols and pentanoic acid (0.6 M) as substrates; **b** varying carbon numbers of fatty acids and 1-pentanol (0.6 M)

K_m *value for water*. The kinetic study on indoxyl acetate hydrolysis with PEG_2-lipase from *P. fluorescens* was performed in a transparent benzene solution in order to obtain the K_m value for H_2O [40]. The double reciprocal plots of the rate (v^{-1}) and the water concentration (S^{-1}) at fixed concentrations of indoxyl acetate gave parallel straight lines, indicating that the hydrolysis takes place as a double-displacement reaction (Ping-Pong) [41]. The K_m value was calculated to be 7×10^{-2} M for H_2O, and 1.6×10^{-1} M for indoxyl acetate, and V_{max} was 4 700 μmol min^{-1} mg^{-1}.

5 Other Conjugates

Furthermore, some bioactive compounds such as coenzymes, natural pigments and drugs have also been coupled with PEG derivatives. For example, PEG-hemin became soluble in an aqueous solution and also in organic solvents and catalyzed peroxidase-like reactions [42]. Nicotinamide adenine dinucleotide and nicotinamide adenine dinucleotide phosphate were also coupled with PEG [43], and the PEG-modified coenzymes could be utilized in bioreactor systems.

5.1 Magnetic Urokinase

A fibrinolytic agent, urokinase, has been used clinically for thrombosis and embolism therapy [44]. However, because of its very short half-life and high susceptibility towards its inhibitors, a huge dosage of the enzyme usually has to be administered. This therapy is, therefore, accompanied by the risk of unnecessary bleeding and other side effects. To overcome these drawbacks, selective delivery or targeting of the drug to the affected parts is most desirable. It has been reported that urokinase was attached to an antibody against fibrin [45]. However, there is a possibility that the antibody raised by an experimental animal may elicit adverse immune reaction in a human body. A more attractive and effective method is to make use of magnetic protein drugs for the selective delivery with a magnetic force.

Recently, we have devised a magnetic enzyme, composed of magnetite (Fe_3O_4), PEG derivative and an enzyme, using an activated magnetic modifier. This activated magnetic modifier has reactive carboxyl groups of α,ω-dicarboxymethylpoly(oxyethylene) bound to magnetite, which easily reacts to amino groups in urokinase [46, 47]. The magnetic urokinase labeled with ^{125}I was injected into the tip part of a tail vein of a mouse, while a magnetic force (0–12 000 G) was applied to the root part of the tail as a target site [48]. The distribution of radioactivity of the magnetic urokinase in various parts of the mouse was measured (Table 8). By using magnetic force of 12 000 G, approximately as much as 60% of the magnetic urokinase injected was localized in the tail root. By applying the magnetic force to the heart, approximately 20% of the radioactivity was attracted to the heart.

5.2 PEG$_2$-Melanin

Melanins, black-colored pigments found in living organisms like mammalian hair, skin and eyes, are known to have a photoprotective function in human

Table 8. Localization of magnetic urokinase on target sites by magnetic force in mice

Magnetic force G	Percent distribution (Mean value $\pm$ S.E.M., n = 3–7)						
	Tail root	Tail tip	Liver	Spleen	Kidney	Lung	Heart
Magnetic urokinase							
A. Targeting to Tail root							
0	0.7 ± 0.2	0.6 ± 0.2	32.1 ± 6.9	2.5 ± 0.5	4.5 ± 1.0	50.8 ± 6.5	0.6 ± 0.1
4000	14.2 ± 3.6	2.9 ± 1.3	20.0 ± 3.1	1.5 ± 0.6	3.7 ± 0.7	43.3 ± 8.3	0.4 ± 0.1
8000	32.9 ± 8.3	2.2 ± 0.8	17.9 ± 3.0	1.8 ± 0.8	5.9 ± 2.9	28.7 ± 8.9	0.5 ± 0.1
12000	60.8 ± 7.2	1.5 ± 0.2	8.2 ± 1.7	0.7 ± 0.1	1.6 ± 0.3	20.4 ± 8.3	0.7 ± 0.4
B. Targeting to Heart							
12000	2.0 ± 0.6	2.0 ± 0.5	7.9 ± 1.9	0.3 ± 0.2	1.9 ± 0.6	39.0 ± 5.7	18.2 ± 2.2
Unmodified urokinase							
	1.5 ± 0.4	2.5 ± 1.0	42.1 ± 1.9	1.3 ± 0.2	21.7 ± 0.7	1.5 ± 0.4	0.5 ± 0.2

beings [49, 50]. However, because of their insolubility in an aqueous solution or in organic solvents, little is known about their physicochemical and biological properties. In order to obtain a soluble melanin, we tried to modify melanin with activated PEG_2 [51]. PEG_2-melanin became soluble with a solubility of 12 g per 100 g of water, and it also became soluble in organic solvents such as chloroform, ethanol, pyridine, dioxane, acetone, benzene and dimethylformamide with solubilities ranging from 6 g to 11 g per 100 g of solvent. A solubilized melanin may not only be useful as a cosmetic pigment but may also reveal the physiological roles of melanin in the derma.

5.3 Chlorophyll-Bentonite Conjugate

Chlorophyll molecules bind to proteins in chloroplasts and exhibit physiological functions. But chlorophylls extracted with organic solvents are quite unstable against light and lose their function. With the aim of making chlorophylls photostable, a clay mineral of montmorillonite (bentonite) was used to prepare chlorophyll a-bentonite conjugate [52]: chlorophyll a dissolved in benzene was adsorbed by lyophilized bentonite powder. The absorption spectrum of the chlorophyll a-bentonite conjugate was measured by the opal glass method [53] after the addition of a small amount of water to the conjugate to form a paste (Table 9). The peak position shifted to a longer wavelength by increasing the amount of chlorophyll a bound to bentonite. The conjugate composed of 4.4 mg chlorophyll and 8.86 mg of bentonite showed the peak position at 678 nm, which is in good agreement with that of intact spinach leaves [53].

Figure 10 shows the stability of chlorophylls from spinach leaves conjugated with bentonite and chlorophylls free from bentonite by irradiation of visible light. No bleaching of the chlorophylls conjugated with bentonite took place even after 40 h of irradiation, although bleaching of chlorophylls proceeded markedly with the irradiation.

Table 9. Wavelengths of absorption maxima of chlorophyll a-bentonite conjugates

Adsorbent (mg)	Chlorophyll a (mg)	Wavelength (nm)
8.86	0.27	671
8.86	0.70	673
8.86	1.3	674
8.86	2.3	675
8.86	4.4	678
Chlorophyll a in benzene		660
Chlorophylls in leaves		678

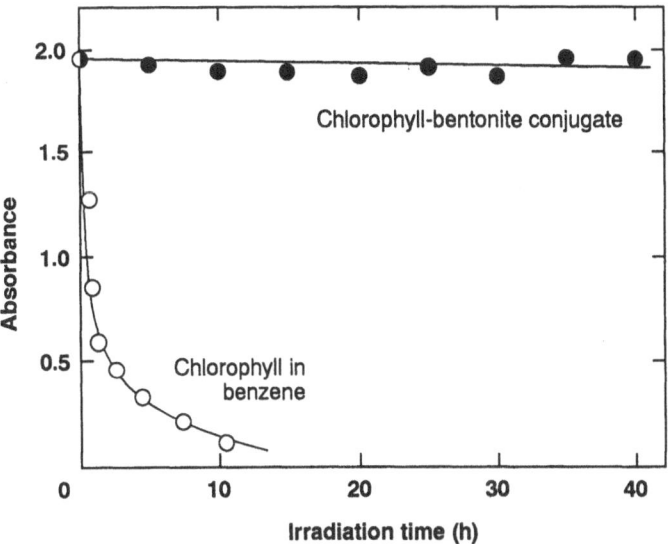

Fig. 10. Photostability of the chlorophyll-bentonite conjugate and chlorophylls free from bentonite. Chlorophylls from spinach leaves were irradiated with visible light ($2100 \, W \, m^{-2}$). The photostability of free chlorophyll and chlorophyll-bentonite conjugate was determined by measuring the absorbance at 660 nm and 673 nm, respectively

References

1. Inada Y, Matsushima A, Kodera Y, Nishimura H (1990) J Bioact Compat Polym 5: 343
2. Inada Y, Kodera Y, Matsushima A, Nishimura H (1992) Enzyme modification with synthetic polymers. In: Imanishi Y (ed) Synthesis of biocomposite materials. CRC Press, London, p 85
3. Curme GO, Johnston F (1952) Glycols. Reinhold, New York
4. Harris JM (1985) Macromol Chem Phys C25: 325
5. Matsushima A, Nishimura H, Ashihara Y, Yokota Y, Inada Y (1980) Chem Lett 773
6. Ono K, Kai Y, Maeda H, Samizo F, Sakurai K, Nishimura H, Inada Y (1991) J Biomater Sci, Polymer Edn 2: 61
7. Abuchowski A, Kazo GM, Verhoest CR, van Es T, Kafkewitz D, Nucci ML, Viau AT, Davis FF (1984) Cancer Biochem Biophys 7: 175
8. Abuchowski A, McCoy JR, Palczuk NC, van Es T, Davis FF (1977) J Biol Chem 252: 3582
9. Carpenter CP, Woodside MD, Kinkead ER, King JM, Sullivan LJ (1971) Toxicol Appl Pharmacol 18: 35
10. Capizzi RL (1981) Cancer Treat Rep 65(suppl. 4): 115
11. Ashihara Y, Kono T, Yamazaki S, Inada Y (1978) Biochem Biophys Res Commun 83: 385
12. Yoshimoto T, Nishimura H, Saito Y, Sakurai K, Kamisaki Y, Wada H, Sako M, Tsujino G, Inada Y (1986) Jpn J Cancer Res 77: 1264
13. Wada H, Imamura I, Sako M, Katagiri S, Tarui S, Nishimura H, Inada Y (1990) Ann N Y Acad Sci 613: 95
14. Yoshimoto T, Ritani A, Ohwada K, Takahashi K, Kodera Y, Matsushima A, Saito Y, Inada Y (1987) Biochem Biophys Res Commun 148: 876
15. Kodera Y, Tanaka H, Matsushima A, Inada Y (1992) Biochem Biophys Res Commun 184: 144
16. Itai A, Yonei M, Mitsui Y, Iitaka Y (1976) J Mol Biol 105: 321

17. Huggins C, Lapides J (1947) J Biol Chem 170: 467
18. Abuchowski A, van Es T, Palczuk NC, Davis FF (1977) J Biol Chem 252: 3578
19. Whitesides GM, Wong C-H (1985) Angew Chem, Int Ed Engl 24: 617
20. Klibanov AM (1990) Acc Chem Res 23: 114
21. Okahata Y, Ijiro K (1988) J Chem Soc, Chem Commun 1392
22. Inada Y, Yoshimoto T, Matsushima A, Saito Y (1986) Trends Biotechnol 4: 68
23. Inada Y, Takahashi K, Yoshimoto T, Ajima A, Matsushima A, Saito Y (1986) Trends Biotechnol 4: 190
24. Inada Y, Takahashi K, Yoshimoto T, Kodera Y, Matsushima A, Saito Y (1988) Trends Biotechnol 6: 131
25. Takahashi K, Ajima A, Yoshimoto T, Inada Y (1984) Biochem Biophys Res Commun 125: 761
26. Takahashi K, Nishimura H, Yoshimoto T, Saito Y, Inada Y (1984) Biochem Biophys Res Commun 121: 261
27. Takahashi K, Yoshimoto T, Ajima A, Tamaura Y, Inada Y (1984) Enzyme 32: 235
28. Lee H, Takahashi K, Kodera Y, Ohwada K, Tsuzuki T, Matsushima A, Inada Y (1988) Biotechnol Lett 10: 403
29. Matsushima A, Okada M, Inada Y (1984) FEBS Lett 178: 275
30. Hiroto M, Matsushima A, Kodera Y, Shibata Y, Inada Y (1992) Biotechnol Lett 14: 559
31. Takahashi K, Ajima A, Yoshimoto T, Okada M, Matsushima A, Tamaura Y, Inada Y (1985) J Org Chem 50: 3414
32. Takahashi K, Nishimura H, Yoshimoto T, Okada M, Ajima A, Matsushima A, Tamaura Y, Saito Y, Inada Y (1984) Biotechnol Lett 6: 765
33. Yoshimoto T, Takahashi K, Nishimura H, Ajima A, Tamaura Y, Inada Y (1984) Biotechnol Lett 6: 337
34. Ajima A, Takahashi K, Matsushima A, Saito Y, Inada Y (1986) Biotechnol Lett 8: 547
35. Takahashi K, Kodera Y, Yoshimoto T, Ajima A, Matsushima A, Inada Y (1985) Biochem Biophys Res Commun 131: 532
36. Matsushima A, Kodera Y, Takahashi K, Saito Y, Inada Y (1986) Biotechnol Lett 8: 73
37. Nishio T, Takahashi K, Tsuzuki T, Yoshimoto T, Kodera Y, Matsushima A, Saito Y, Inada Y (1988) J Biotechnol 8: 39
38. Kodera Y, Takahashi K, Nishimura H, Matsushima A, Saito Y, Inada Y (1986) Biotechnol Lett 8: 881
39. Kikkawa S, Takahashi K, Katada T, Inada Y (1989) Biochem Int 19: 1125
40. Matsushima A, Okada M, Takahashi K, Yoshimoto T, Inada Y (1985) Biochem Int 11: 551
41. Cleland WW (1963) Biochim Biophys Acta 67: 104
42. Takahashi K, Matsushima A, Saito Y, Inada Y (1986) Biochem Biophys Res Commun 138: 283
43. Hayakawa K, Urabe I, Okada H (1985) J Ferment Technol 63: 245
44. Ploug J, Kjeldgaard NO (1957) Biochim Biophys Acta 24: 278
45. Bode C, Matsueda GR, Hui KY, Haber E (1985) Science 229: 765
46. Inada Y, Ohwada K, Yoshimoto T, Kojima S, Takahashi K, Kodera Y, Matsushima A, Saito Y (1987) Biochem Biophys Res Commun 148: 392
47. Yoshimoto T, Ohwada K, Takahashi K, Matsushima A, Saito Y, Inada Y (1988) Biochem Biophys Res Commun 152: 739
48. Yoshimoto T, Saito Y, Sugibayashi K, Morimoto Y, Tsukada T, Kanmatsuse K, Kodera Y, Matsushima A, Inada Y (1989) Drug Delivery System 4: 121
49. Wick MM (1986) Melanin: structure and properties. In: Fitzpatrick TB, et al. (ed) Brown Melanoderma: Biology and Disease of Epidermal Pigmentation Univ Tokyo Press Tokyo p 37
50. Nicolaus RA (1968) Melanins Hermann Paris
51. Ishii A, Furukawa M, Matsushima A, Kodera Y, Yamada A, Inada Y, Dyes Pigments (in press)
52. Kodera Y, Kageyama H, Sekine H, Inada Y (1992) Biotechnol Lett 14: 119
53. Shibata K (1959) Methods Biochem Anal 7: 77
54. Kamisaki Y, Wada H, Yagura T, Matsushima A, Inada Y (1981) J Pharmacol Exp Therap 216: 410
55. Sasaki H, Ohtake Y, Matsushima A, Hiroto M, Kodera Y, Inada Y (1993) Biochem Biophys Res Commun 197: 287

Breakdown of Plastics and Polymers by Microorganisms

Fusako Kawai
Department of Biology, Kobe University of Commerce, Kobe 651-21, Japan

The interest in environmental issues is still growing and there are increasing demands to develop materials which do not burden the environment significantly. Awareness of the waste problem and its impact on the environment has awakened new interest in the area of degradable polymers. Biodegradation is necessary for water-soluble or water-miscible polymers because they eventually enter streams which can neither be recycled nor incinerated. It is important to consider the microbial degradation of natural and synthetic polymers in order to understand what is necessary for biodegradation and the mechanisms involved. This requires both biochemical insight and understanding of the interactions between materials and microorganisms. It is now widely requested that polymeric materials come from renewable resources instead of petrochemical sources. The microbial production of polymeric and oligomeric materials is also described.

Advances in Biochemical Engineering/
Biotechnology, Vol. 52
Managing Editor: A. Fiechter
© Springer-Verlag Berlin Heidelberg 1995

List of Abbreviations

Tm: melting point
Tg: glass-transition temperature
HPLC: high-performance liquid chromatography
LC-MS: liquid chromatography-mass spectrometry
Mn: average molecular weight
DCIP: 2,6-dichlorophenolindophenol
MTT: 3-(4,5-dimethyl-2-thiazolyl)-2,5-diphenyl-2H-tetrazolium bromide
PMS: phenazine methosulfate
p-CMB: *p*-(chloromercuri)benzoic acid

1 Introduction

This review summarizes what we know about the biodegradation of natural and synthetic polymers by microorganisms. Natural polymers are inherently biodegradable, but biodegradability of chemically modified polymers varies, depending on the extent of modification. Synthetic polymers are divided into two types, water-soluble and water-insoluble. The former are generally specialty polymers with functional groups that effect water solubility such as carboxyl, hydroxyl, amido, etc.; the latter are usually nonfunctional polymers commonly referred to as commodity plastics. Commodity plastics are used in packaging, disposable diaper backing, fishing nets, and agricultural film. They include polymers such as polyethylene, polypropylene, polystyrene, poly(vinyl chloride), poly(ethylene terephthalate), and Nylon 6,6. Some synthetic polymers are biodegradable, although most commodity plastics used now are non-biodegradable.

Plastics have reached a level of over 10 weight percent and 30 volume percent of solid waste at present in the USA and Japan, a large proportion of which are commodity plastics. Their lack of degradability is impacting significantly on the rate of depletion of landfill sites and their persistence adds to the growing water and land surface litter problem. This has raised growing concern about degradable polymers and promoted research activity worldwide to either modify current products to promote degradability or to develop new alternatives that are degradable by any or all of the following mechanisms: biodegradation, photodegradation, environmental erosion and chemical degradation. Biodegradation may lead to complete removal of the material from the environment. On the other hand, specialty polymers represent a wide range of applications for generally low volume products. They are used in cosmetics, water treatment, and as dispersants, thickeners, detergents, and super absorbents; they include poly(acrylic acid), polyacrylamide, poly(vinyl alcohol), and poly(ethylene glycol). Some enter the environment as landfill, along with commodities, e.g. poly(acrylic acid) as super absorbents in diapers, poly(acrylic acid) and poly(alkylene oxides) in detergent effluents and others from industrial effluents, e.g. poly(vinyl alcohol) from paper and textile mills. Water-soluble polymers entering the environment do not cause a litter problem and are, therefore, of little concern. However they enter streams, and even in buried landfill sites they eventually pollute the ground and surface waters and can be neither incinerated nor recycled. Therefore, biodegradability is a very cogent issue. Medical polymers are a unique case of specialty polymers. These have been designed to degrade mainly by hydrolysis in vivo where they perform such functions as controlling drug release and providing temporary supports such as sutures in post-operation recovery. They are widely accepted and used. Commodity plastics and specialty polymers should either have predictable lifetimes and degrade on exposure to the chosen release environment, or be nondegradable and capable of being recycled or incinerated.

Over the last two decades, much research has been done on the microbial degradation of synthetic and natural polymers. These studies have resulted in general guidelines concerning the relationship between structure and biodegradability, which is introduced in this review and will contribute to the development of biodegradable polymers.

2 Natural Polymers

Natural polymers or polymers based on naturally derived monomers offer the biggest incentive for future development. They are products from renewable resources that totally biodegrade in their natural form, given the choice of environment. Natural polymers which are components of living plants, animals and algae have a long history as paper (cellulose), foods (starches, proteins), sizing materials and gelatinizers (starches and algal materials), etc. Their recently expanded usage as commodity plastics (starches, proteins, cellulose etc.), food additives (cellulose and algal materials) and in medical applications (chitin and chitosan) depends on the modification of their chemical and physical properties by blending or copolymerizing with other biodegradable polymers or by chemical modification. In addition, microbial production of some polymers and monomers has been reported. Polyhydroxyalkanoic acids and polysaccharides are representative polymeric materials produced by bioprocesses. Amino acids, and organic acids, such as lactic acid and malic acid, are microbial products used as monomers for synthesizing polymeric materials.

2.1 Naturally Occurring Polymers

Cellulose, lignin, gluten, zein, chitin, chitosan, starch and alginic acid are included in this category. Celluloses from various origins are used as papers, but they are also chemically modified and used for various other purposes. Biodegradability of modified cellulose is highly dependent on the degree of substitution (DS) and the length of the acyl group. In general, as the DS and the length of the acyl group decreases, the rate of biodegradation increases. Cellulose acetate [1] (1.7- and 2.5-DS), powder or film, is sufficiently mineralized to CO_2 under aerobic thermophilic conditions (compost environment); conversions of greater than 70% of cellulose acetate (1.7- and 2.5-DS) substrates have been measured as recovered CO_2. Buchanan [2] suggested that cellulose acetates having a DS less than approximately 2.1 completely disintegrated after approximately 15 days, a composing rate comparable to that of other commercial biopolymers such as polyhydroxybutyrate. As a thermoplastic, these low DS cellulose acetates have thermal and physical properties identical in many respects to commercial cellulose acetate. These properties are high flexural modulus and strength, relatively high melting temperature, and a narrow Tm–Tg gap.

Lignin is the most recalcitrant natural product. White-rot fungi (basidio-mycetes) play an important roles degrading lignin in nature, but biodegrada-tion rates are very slow. Biodegradation of lignin has been mainly studied with model compounds, since the natural material is degraded with difficulty and a vast amount is yielded as a by-product of cellulose production in woods. Hatakeyama et al. [3] chemically synthesized polyurethane sheets and foams from lignin, molasses, cellulose powder, wood meal and coffee grinds dissolved in polyethylene glycol which showed superior biodegradability. Saccharides and lignin residues acted as hard segments in polyurethanes.

Natural polyamino acids are very important resources which have been used widely and their application is expected to expand to biodegradable plastics. These included gluten and zein from cereals, collagen from animal tissues, silk fibroin and polyglutamic acid produced by bioprocesses. Yasui et al. [4] reported that gluten mixed with glycerine or diethylene glycol could be moulded into plastics which were completely degraded by 4 weeks in a soil burial test. Jane et al. [5] reported that biodegradable plastics can be formed from starch-zein blends. Since silk is an expensive fabric material, application of waste silk produced during fabric processing is being studied. Silk protein, fibroin, extracted from silk has been suggested as a carrier for controlled drug-release and as a support for en-zymes in biochemical applications. Collagen is being used as sausage casing, for cosmetics, and also medically applied as sutures and artificial skin. Polyglutamic acid is known as an extracellular polysaccharide of "natto", which is a traditional Japanese biotreated food produced by *Bacillus natto* from steamed soy beans. Microbial production and application of this material is described in Section 2.2.

Chitin is a homopolymer composed of N-acetylglucosamine which biosynthe-sized at an annual level of 10^{10}–10^{11} tons corresponding to that of cellulose. A large portion of chitin is wasted in contrast to the extensive utilization of cellu-lose. Although it will be a new natural resource in the future, its application to commodity plastics has not been reported, because of insolubility and infusibil-ity. Chitin is a component of shells and crusts of *Crustacea* and insects and is industrially produced from waste crab shells from the fishing industry. Chitosan is a highly deacetylated chitin, occurring naturally in several fungi. Chitosan is usually produced by chemical deacetylation of chitin as a water-soluble poly-mer which can be chemically modified. Chitin and chitosan at present are used for water-treatment, and as cosmetics, foods and the medical field. Chemically-modified chitosan is used as an ingredient in cosmetics, drug delivery systems and pharmaceuticals. It forms microfibrils and may be used for numerous applica-tions, the majority of which are still in the developing stage. Major obstacles for large-scale technical application of high quality chitin/chitosan are the high price of the polymers and the difficulties in preparing uniformly reproducible charges in bulk quantities from various marine organisms.

Starches are the most acceptable renewable materials used for commodity plas-tics because of their abundance and low cost. Many studies have been done with degradable starch-blends (see Sect. 3.7.).

Alginic acid and agar are polymers extracted from brown and red algae. Alginic acid is a linear polyuronic acid composed of poly(β-D-mannuronic acid) and poly(α-L-glucuronic acid) which can be used as paper, non-woven cloth and transparent sheets. This polymer is degraded mainly by lyases of bacteria, molds, brown algae, and marine *Mollusca* and also by the hydrolase from *Alginovibrio aquatilis* [6]. Alginic acid lyase may be applicable in cystic fibrosis, a genetic disease involving *Pseudomonas aeruginosa*. Interestingly mucoid strains of *P. aeruginosa* synthesize an intracellular polymannuronide lyase and nonmucoid forms of the organism produce much lower amounts of the enzyme [7]. Agar is a well-known polysaccharide (Ca-Mg salts of galactan sulfate esters) which is widely utilized as solidifier for foods and culture media. There are no reports on industrial application of this material as alternatives for water-soluble synthetic polymers or commodity plastics.

2.2 Microbial Production of Polymers and Monomers

It is a general rule that microbial products can be degraded by microorganisms. One of the most exciting developments are polymers based on bacterially produced polyhydroxyalkanoate (PHA) shown in Fig. 1. In 1925, poly-(3-hydroxybutyrate), P(3HB), was first discovered by Lemoigne as a bacterial storage material during periods of nutrient stress [8]. The content of polyester in dried cells amounted up to 80 weight %. Since this material has high crystallinity and weak mechanical strength and is not suited for practical application to thermoplastics, the bacterial production of a variety of PHAs was examined with many bacteria from various carbon sources like sugars, alkanoic acids and alcohols (Table 1). More than 40 hydroxyalkanoates were made into polymers. ICI in Great Britain started small-scale production of P(3HB-*co*-3HV) (3-hydroxyvalerate (3HV) content: 0–40 mol %) under the commercial name of BIOPOL [9]. The present production level is a few hundred tons per year [10]. This polymer has physical properties similar to those of polypropylene. Doi et al. also developed efficient production of this copolymer (3HV content; 0–95 mol %) by *Alcaligenes eutrophus* [11]. The production of other copolymers has been suggested by several research groups: P(3HB-*co*-4HB) (4-hydroxybutyrate (4HB) content; 0–100 mol %) was produced with *A. eutrophus* from various carbon sources; 4HB, 1,4-buranediol, 1,6-hexane diol and butyrolactone [12, 13]. Also P(3HB-*co*-3HP) (3HP: 3-hydroxypropionate) was biosynthesized with *A. eutrophus*, *A.*

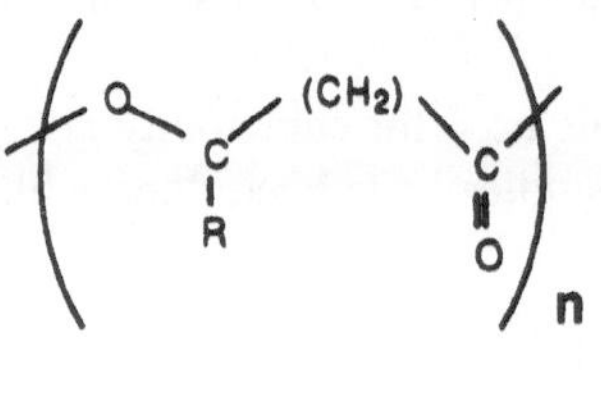

Fig. 1. Structure of PHA. R = alkyl, haloalkyl, nitrile-substituted alkyl, and aryl

Table 1. Bacterial production of a variety of PHAs

Polyester produced	Bacteria
P(3HB)	Many strains
P(3HB-co-3HV)	Alcaligenes eutrophus
	Bacillus megaterium
	Chromatium vinosum
	Rhodobacter sphaeroides
	Pseudomonas cepacia
	Methylobacterium extorquens
P(3HB-co-4HB)	Alcaligenes eutrophus
	Alcaligenes latus
	Comamonas acidovorans
P(3HA)	Pseudomonas putida
3HA: C_3-C_{11}	Pseudomonas aeruginosa
	Pseudomonas oleovorans

latus and other bacteria [14] and P(3HA), in which long chain side groups are included, was formed with *Pseudomonas oleovorans*. [15]. The physical and mechanical properties of the copolymers were strongly dependent upon their composition. These copolyesters behave as thermoplastics and can be processed into transparent film and strong fiber by conventional extrusion and moulding equipment. They are biodegradable and biocompatible and the promising biodegradable plastics in the future, although they are presently expensive, and the question of commercial viability has not yet been answered. Ramsay [16] examined a P(3HB-co-3HV) blend with inexpensive starch to reduce the cost of the film; it showed superior biodegradability and practical-level thermoplasticity. Chemical synthesis of PHAs has also been examined.

Biosynthesis of PHAs are catalyzed by PHA synthetase via hydroxyalkanoyl coenzyme A. Genes involved in PHA synthesis by *A. eutrophus* etc. have been cloned by many researchers. Recently, Poirier et al. cloned the genes of *A. eutrophus* into a plant, *Arabidopsis thaliana* [17].

Just as bacterially produced PHAs appear to be the best route to new biodegradable plastics, bacterial celluloses, polysaccharides, etc., may represent new water-soluble polymers. A recent book edited by Byrom [18] points out the value of bacterial polymers which include, in addition to the PHAs, cellulose, hyaluronic acid, and alginates. Some will be expensive, but through a combined effort of biochemistry and polymer science, new polymers will be developed through biological synthesis and chemical modification of these natural polymers. These could meet the demands for environmentally biodegradable polymers. This area will become a major focus for research and development activities over the next decade [19]. Of the available microbial polysaccharides [20] the most important products are dextran made by *Leuconostoc mesenteroides* and xanthan gum by the plant pathogen, *Xanthomonas campestris*. Industrially important polysaccharides are summarized in Table 2. They are used in medical fields as blood fillers, low calory materials, carriers for metallic ions and drugs.

Table 2. Representative microbial production of polysaccharides

Polysaccharide produced	Microorganism
Dextran	*Leuconostoc mesenteroides*
Pullulan	*Aureobasidium pullulans*
Cellulose	*Acetobacter xylinum, Ac. pastourianus, Ac. hansenii* etc.
Curdlan	*Alcaligenes faecalis* var. *myxogenes*, *Agrobacterium, Rhizobium*
Scleroglucan	*Sclerotium*
Schizophylan	*Schizophyllum*
Succinoglucan	*Alcaligenes faecalis* var. *myxogenes*, *Agrobacterium, Rhizobium*
Levan	*Acetobacter levanicum, Bacillus subtilis*
Xanthangum	*Xanthomonas campestris, X. oryzae, X. phaseoli* etc.
Arthrobacter polysaccharide	*Arthrobacter viscosus, A. stabilis, A. globiformis* etc.
Alginic acid	*Pseudomonas aeruginosa, Azotobacter vinelandii*

This area has recently attracted attention since some polysaccharides have shown antitumor activity or immune-enhancing activity (kurestin, lentinan, schizophyllan, etc.). They are also used in biochemical research as supports for enzymes, gel filtration, and affinity chromatography. Most of them are very water-soluble and not useful as bioplastics. Several polysaccharides are utilized as food stabilizers, gluing agents for oil production and paints, coating and gelification agents for foods, food additives, sizing agents for fabrics and papers, fillers for ice cream and jelly, and as water-treatment agents. Some show pseudoplastic properties (scleroglucan), and form film and fibre (pullulan), and heat-curdling gels (curdlan), crystalline fibrils (bacterial cellulose) and films (xanthan and *Arthrobacter* polysaccharide). These polymeric materials can expand their scope to new application areas.

Polyglutamic acid (γ-PGA) and polylysine (ε-PL) are polyamino acids which are microbiologically synthesized. They differ from natural and synthetic polyamino acids which are all α-polyamino acids (α-PAA), as shown in Fig. 2. γ-PGA was first discovered as a capsular component of *Bacillus anthracis* in 1973 [21]. Later, γ-PGA was detected in a culture filtrate of *Bacillus subtilis* by Bovarnick [22] and also found as a component of a traditional Japanese food, "natto" (*Bacillus natto*) [23]. The material obtained as a capsular component of *B. anthracis* was the D-homopolymer. The polymer produced in the culture filtrate is a mixture of D- and L-homopolymers (Mn: 100 000–1 000 000), the biosynthesis of which was discussed by Hara et al. [24]. The productivity of the polymer is influenced by cultural conditions and the content of L-homopolymer reached more than 95% under optimum conditions. γ-PGA-producing bacteria were identified as *Bacillus* spp.: *B. subtilis, B. lichiniformis, B. anthracis, B. megaterium, B. natto* and *B. subtilis* var. *polyglutamicum*. The fiber from L-glutamic acid was more than 30 g per liter [25]. This polymer can be degraded by animal γ-glutamic acid transpeptidase [26], a bacterial enzyme from the polymer degrader, *Flavobacterium polyglutamicum* [27], and also by an extracellular

$$R$$
$$|$$
-CO-(-NH-CH-CO-)$_n$-NH α -PAA

$$COOH$$
$$|$$
-(-NH-CH-(CH$_2$)$_2$-CO-)$_n$- γ -PGA

$$NH_2$$
$$|$$
-(-NH-(CH$_2$)$_4$-CH-CO-)$_n$- ε -PL   Fig. 2. Structures of polyamino acids

degrading enzyme produced in the late growth phase of a PGA-producer, *Bacillus subtilis* ATCC9945 T (later reidentified as *B. lichiniformisT*) [28], but hardly by ordinary proteinases. The material is not yet commercially available, but is expected to be applied as a thickener for foods, and in medical drugs, pesticides, cosmetics, paints, ceramics, as a carrier for controlled drug-release, support for enzymes, binder and a humid insulator.

ε-PL is another microbial polyamino acid which is a homopolymer of L-lysine ($n = 25$–30) and is produced by *Streptomyces albulus* [29]. The polymer has antibacterial activity [30] and is already on the market as a new food preservation agent. Expansion of its application is expected as an antibacterial agent and cationic polymer additive to cosmetics, toiletary goods, medical drugs, and agricultural chemicals.

Poly(L-malic acid) (Mn: 5000) was first detected in a water extract of a wheat bran culture of *Penicillium cyclopium* as a protease inhibitor [31]. *Aureobasidium* sp. A-91 accumulated approximately 60 g of poly(β-L-malic acid)-Ca salt in 1 l of culture grown on glucose [32], the molecular weight of which was 6 000–11 000. This material has pendant carboxyl groups and can be used as a bioabsorbable polymer in the medical field. Poly(α and β-L-malic acid) and poly(α,β-L-malic acid) are also chemically synthesized for use as useful bioabsorbable polymers [33–35]; this will be described in Sect. 3.2.

Monomers produced from natural resources represent another opportunity to be exploited in the search for biodegradable polymers. Polylactic acid (PLA) is at present mainly used as a bioabsorbable polymer applied in medicine for sutures, controlled-release devices, internal bone fracture fixation and bone reconstruction. Currently, there is increasing interest in using lactic acid-based polymeric compounds for disposable and degradable plastic materials. Already, several major manufacturers are contemplating production of PLA from cheaply available monomer produced by fermentation for use in packing film and other industrial applications, in addition to its current use in medical applications. This material is chemically synthesized from lactic acid which is a bio-product. The main limitations to actual use are processing difficulties and production costs. Argonne National Laboratories have fine-tuned a process for making lactic acid from potato and cheese waste; the technique cuts the time for converting polysaccharide to sugars from 100 h to less than 10 h and includes the synthesis of polymer from lactic acid [36]. Recently Ishizaki et al. [37] reported the production of L-lactic acid from xylose, a major component of lignocellulose which is an agricultural waste, by *Lactococcus lactis* with the high yield of 47%. It is worth noting that chemical and processing engineering has not yet been largely applied as was the case when synthetic commodity polymers were developed many years ago. Whether the natural polymers can be biodegraded by environmental microorganisms has not been shown yet, although enzymatic hydrolysis and non-enzymatic hydrolyses have been suggested and degradation by-products can be nutrients for microorganisms.

Polyglycolic acid (PGA) is another bioabsorbable polymer often used as PLA-PGA blends and copolymers; Dupont began to supply PGA, PLA and copoly(GA/LA) under the name of "Medisorb" [38]. In addition, polymalic acid (PMA) is attracting increasing attention as a new absorbable polymer. These poly(alkanoates) are of the poly(α-hydroxyacid) type and chemically degrade in animals followed by biological assimilation or mineralization, in contrast to poly(alkanoates) of the poly(β-hydroxy acid)-type (PHB, PHB/PHV) which appear quite resistant to degradation in animals. In general, PLA is degraded in 3–6 months, PGA in 2–4 weeks and PMA in 3–5 days. Glycolic acid and malic acid can be biologically or enzymatically produced from various carbon substrates.

Various kinds of amino acids are produced by fermentation technology. Several polyamino acids such as polyglutamic acid and polyaspartic acid and their copolymers are chemically synthesized from their corresponding monomer amino acids and used mainly as drug-delivery systems and other medical materials [39]. These chemically-synthesized polyamino acids are α-polyamino acids, while microbial polyamino acids like polyglutamic acid and polylysine are γ- and ε-polypeptides, respectively. Natural proteins and microbial polyamino acids are superior in biodegradability, but they are expensive, have limited productivity and there is difficulty getting uniform quality which depends upon their molecular size. Chemically synthesized polyamino acids have uniform qualities and a variety of designs. The first synthetic product was industrially produced in 1976 in Japan as a surface-treatment agent for synthetic skin. Although they show high biodegradability, the polymers lack heat plasticity and cannot be utilized as plastics. Recently the Donlar Corp. developed a new condensation synthesis of polyaspartic acid [40] as an alternative to polyacrylic acid which is widely used as a functional polymer, but has low biodegradability. This polyanionic peptide is a copolymer having approximately 70% of the amide bonds formed from the β-carboxyl, rather than exclusively from the α-carboxyl. Upon acid hydrolysis, a racemic mixture of aspartic acid is produced, rather than exclusively L-aspartic acid. Wheeler et al. found susceptibility of the polymer to proteolytic fragmentation by effluents and activated sludge from a waste water treatment plant [41], although a small amount of potentially persistent metabolites was formed.

3 Biodegradable Synthetic Polymers and Oligomers

Intensive studies on biodegradation of synthetic polymers and oligomers has clearly shown the close correlation between biodegradability and chemical structure. The synthetic polymers that biodegrade tend to have structures similar to those found in naturally occurring polymers, suggesting that microbial populations produce enzymes that do not discriminate between polymers of similar structure. This is a good indicator for future research directions. Examples of synthetic polymers that biodegrade include poly(vinyl alcohol), which is the only carbon-chain polymer to biodegrade, poly(ethylene oxide), poly(lactic acid), aliphatic polyesters, polycaprolactone, and polyamides. Several oligomeric structures which biodegrade are known: oligomeric ethylene, styrene, isoprene, butadiene, acrylonitrile, and acrylate. These results suggest that hydrolysis and oxidation are the primary processes for polymer degradations. Among the important factors which contribute to the biodegradability of polymers are the presence of hydrolyzable and oxidizable groups, balance of hydrophobicity and hydrophilicity, and optical specificity. Physical properties such as crystallinity, orientation and morphological properties such as surface area, affect the rate of degradation [42].

3.1 Vinyl Polymers

Polyethylene (PE) is used for a large proportion of commodity plastics, but its biodegradability is very poor. It is known that PE will not degrade to low-molecular weight metabolites in the time span of human life [43, 44]; in other words, PE is regarded as a chemically "inert" polymer, due to its long degradation time. Albertsson et al. showed that PE is biodegraded after a long period; this can only be detected by using labelled materials and liquid scintillation counting [45]. Degradation rates are influenced by factors such as irradiation, morphology, surface area, antioxidants, additives, and molecular weight [46]. Scott had already concluded in 1975 that attack by microorganisms is a secondary process [47]. The step which determines the rate at which degradable polyethylene is returned to the biological cycle appears to be the rate of molecular oxidation which reduces the molecular weight of the polymer to the value required for biodegradation. Even in the absence of any biodegradative attack, the carboxylic acids produced are ultimately oxidized to carbon dioxide and water. An improved mechanism for the biodegradation of polyethylene was recently presented. It proposes that PE is catabolized in a manner similar to that of fatty acids and paraffins in man and animal [45]; initially, an abiotic step in the oxidation of the polymer chain is necessary. Once hydroperoxides have been introduced, a gradual increase in the number of keto groups in the polymer is observed, which is followed by a decrease in keto groups when short-chain carboxylic acids are released to the surroundings as degradation products. In this process, PE chains are degraded stepwise at their chain ends, resulting in a decrease in chain length by two methylene units at a time.

It is generally accepted that PE, when its molecular weight is lower than approximately 1000, can be biodegraded. Approaches towards biodegradable PE include introduction of a photodegradable group into the polymer chain and addition of photosensitized reagents which may be cleaved. Copolymerization with small amounts ($\sim 1\%$) of carbon monoxide or a vinyl ketone provide a polymer which photodegrades at a controlled rate when exposed to the UV radiation in natural sunlight [48]. The Shell organization recently commercialized polyketone($-COCH_2CH_2$)$_n$ (brand name: Carilon) [49]. PE-starch blends which have been reported by several investigators, will be discussed in Sect 3.7.

Microbial degradation of the following vinyl oligomers has been suggested: polystyrene ($n = 2$) [50], polybutadiene ($n = 4$–10) [51], acrylonitrile trimer [52], polyisoprene (cis-1,4-type, $n = 2$–15) [53] and polyacrylate (Mn: ~ 3000) [54–59].

Polyacrylate (PA) is currently used in large quantities as a functional polymer; cross-linked high molecular weight PAs are used as super absorbents and low molecular weight PAs are used as detergents, dispersants, or as water-treatment chemicals. Biodegradation of PA has been reported by activated sludges [54], by oligomer-utilizing bacteria [55–58] and by methanogenic biofilm reactors [59]. Kawai [55, 56] isolated three kinds of bacteria (*Microbacterium* sp., *Xanthomonas maltophilia* and *Acinetobacter* sp.) which assimilated 1,3,5-pentane tricarboxylic acid (PTCA) as sole carbon and energy source. Her intact cells degraded PA of molecular weight 1000–3000. Hayashi et al. isolated acrylic oligomer-utilizing *Arthrobacter* sp. and unidentified bacteria. [57, 58].

PVA is the only carbon-chain polymer to biodegrade at high molecular weights. The biodegradability of this material is greatly dependent on pendant hydroxyl group which lead to water-solubility and susceptibility to oxidation. Microbial degradation of PVA has been reported by several groups. Suzuki et al. isolated *Pseudomonas* 0–3 which assimilated PVA as sole carbon source [60] and added the organism to activated sludge for the treatment of PVA-containing waste waters [61]. Their isolate assimilated PVA up to a polymerization degree of 2000 (saponification, 99.3%). Watanabe et al. isolated PVA-utilizing *Pseudomonas vesicularis* PD [62, 63]. Sakazawa et al. reported that PVA was efficiently degraded by a symbiotic mixed culture of *Pseudomonas putida* VM15A and *Pseudomonas* sp. VM15C [64], where the latter was a PVA degrader and the former supplied PQQ, a growth factor [65]. Both aerobic and anaerobic degradation of PVA have been reported [66, 67].

3.2 Polyesters

Synthesized aliphatic polyesters are generally known to be susceptible to biological attack [68–71]. Tokiwa et al. isolated *Penicillium* spp. 14–3 and 26–1 which utilize polyethylene adipate (PEA, Mn: 3000) and polycaprolactone (PCL, Mn:

25 000) as sole carbon and energy sources, respectively [72, 73]. These fungi assimilate a variety of aliphatic polyesters, but do not grow on alicyclic and aromatic polyesters. An extracellular enzyme from *Penicillium* sp. 14–3 was purified and shown to be a lipase [74]. On the other hand, commercially available lipases from various microorganisms and hog pancreas and hog liver esterases hydrolyze various polyesters [75, 76]: aliphatic and alicyclic polyesters, ester-type polyurethanes, copolyesters composed of aliphatic and aromatic polyesters and copolyamide-esters. PHB, polyglycolide, co(GA/LA, 92/8) and aromatic polyesters were not degraded by these enzymes. Physicochemical properties like melting point, crystallinity, etc. seemed to be closely related to the biodegradabilities of the materials [77].

Recently, an aliphatic polyester with high Mn was commercialized by Showa Denko K.K. and Showa High Polymer Co. Ltd. under the trade name of "Bionolle". It is made of various glycols and dicarboxylic acids. Its properties and biodegradation rates were modified by changing the individual components [78].

Bioabsorbable polymers like PGA, PLA and PMA are known to be nonenzymatically hydrolyzed in vivo; degradation is accelerated by enzymatic hydrolysis. PHAs and other polyesters are also gradually hydrolyzed non-enzymatically in water.

3.3 Polyethers

Polyethers have a long history as specialty polymers used as raw materials for synthesizing detergents or polyurethanes. They are either water-soluble or oily liquids which eventually find their way into environmental or waste water systems. Biodegradability is necessary for materials entering streams because they can neither be recycled nor incinerated. The polyethers include polyethylene glycol (PEG), polypropyleneglycol (PPG), and polytetramethylene glycol (PTMG) (Fig. 3). PEGs are manufactured in large quantities and used as commodity chemicals in various industrial products such as pharmaceuticals, cosmetics, and lubricants. The most common hydrophilic moieties contained in the nonionic surfactants are the ethylene oxide polymers. The majority of PEGs produced are used as nonionic surfactants. PPGs are used in their original form in lubricants, inks and cosmetics, but most are transformed to polyurethanes or surface active agents. PTMG is used exclusively as a constituent of polyurethane. Oligomers up to octamers are water-soluble and can be washed out from polymers with water [81].

Yamagishi and associates at Daicel reported that polyisobutylene oxide may be a promising photodegradable plastic [79]. This material is highly crystalline and exposure of a 25 μm film under UV light at 60 °C leads rapidly to acetone, formic acid and acetic acid as principal degradation products which are readily integrated into the ecological cycle. Monsanto has developed polymerized glyoxylic esters which are polyacetals, stable in alkali, but not in acid. In use as detergents, they

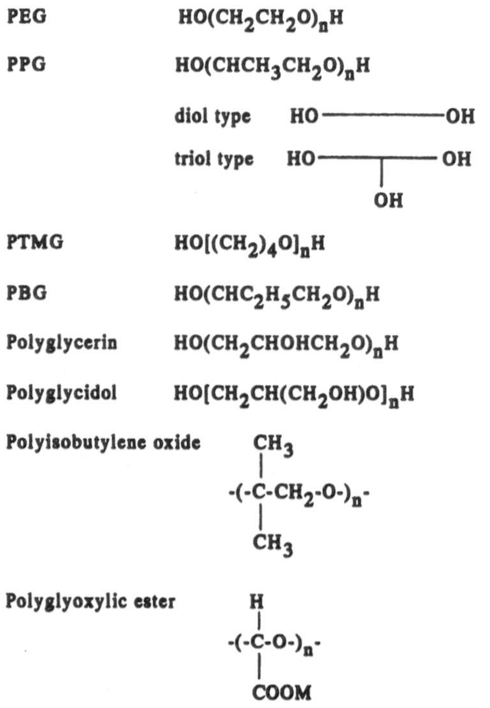

PEG $HO(CH_2CH_2O)_nH$

PPG $HO(CHCH_3CH_2O)_nH$

         diol type   HO————————OH

         triol type   HO————┬———— OH
                            OH

PTMG $HO[(CH_2)_4O]_nH$

PBG $HO(CHC_2H_5CH_2O)_nH$

Polyglycerin $HO(CH_2CHOHCH_2O)_nH$

Polyglycidol $HO[CH_2CH(CH_2OH)O]_nH$

Polyisobutylene oxide

$$-(-\underset{\underset{CH_3}{|}}{\overset{\overset{CH_3}{|}}{C}}-CH_2-O-)_n-$$

Polyglyoxylic ester

$$-(-\underset{\underset{COOM}{|}}{\overset{\overset{H}{|}}{C}}-O-)_n-$$

Fig. 3. Structures of polyethers

are stable, but are hydrolyzed as the pH falls to ~ 7 in sewer systems to the biodegradable monomer [80].

PEG-, PPG-, and PTMG-utilizing bacteria have been isolated from soils and activated sludges [81]. These polyethers were thought to be degraded only by oxidative processes of aerobic bacteria. However, anaerobic degradation of PEG has been studied by a few groups, as described later.

Since the first report on microbial degradation of PEG 400 by Fincher and Payne [82], extensive studies have been done on a series of PEGs which have different average molecular weights, as shown in Table 3. Thus PEGs up to 20 000 are thought to be sufficiently biodegradable.

We isolated various PEG-utilizing bacteria by various enrichments on PEGs 400 to 6000 and classified them into five groups according to their PEG-assimilating limits: PEG 400-, PEG 600-, PEG 1000-, PEG 4000-, and PEG 20 000-utilizing groups. At present, all groups except PEG 600-utilizing bacteria are maintained and have been reidentified [83]. Low PEGs up to 1000 were utilized by various common bacteria. High PEGs from 4000 to 20 000 were utilized by a new genus, *Sphingomonas*. This might suggest that specific membrane structures are necessary for the metabolism of large molecules. PEG 4000-utilizing bacteria were reidentified as a new species of *Sphingomonas*, named *S. macrogoltabidus*. The main component of PEG 20 000-utilizing symbiotic mixed cultures was reidentified as another new species of *Sphingomonas*, named

Table 3. Synopsis of the microbial degradation of PEG compounds

Researcher	Year	Microorganism	PEG compound
Fincher and Payne [13]	1962	diEG-PEG 400	Soil bacterium
Borstlap and Kortland [14]	1967	PEG 400 and 1000	Activated sludge
Patterson et al. [15]	1970	<PEG 400	Air-dried activated sludge
Sturm [16]	1973	<PEG 1000	Acclimated sewage sludge
Pitter [17]	1973	EG-PEG 3500[a]	Sludges and pure culture
Ohmata et al. [18]	1974	triEG-PEG 400	Soil bacterium (possibly *Pseudomonas*)
Harada and Nagashima [19]	1975	triEG or PEG 400	Sludge or soil bacterium (*Alcaligenes*)
Ogata et al. [20]	1975	PEG up to 20 000	Pure cultures and
Kawai et al. [21]	1977		Symbiotic mixed cultures[b]
Haines and Alexander [22]	1975	PEG up to 20 000	Soil bacterium (*Pseudomonas aeruginosa*)
Cox and Conway [23]	1976	PEG up to 4000	Adapted activated sludge
Jones and Watson [24]	1976	EG-PEG 400	Possibly an *Actinetobacter*
Watson and Jones [25]	1977	EG-PEG 1500	*Acinetobacter Pseudomonas Flavobacterium*
Suzuki and Kusunoki [26]	1977	PEG 200–2000	Acclimated activated sludge
Hosoya et al. [27]	1978	PEG 400 and 6000	Soil bacteria
Jenkins and Cook [28]	1979	EG-PEG 200 PEG 200–4000	Bacteria
Thélu et al. [29]	1980	diEG-PEG 400	Soil bacterium (*Pseudomonas*)
Pearce and Heydeman [30]	1980	EG-PEG 1500 (different utilization range)	*Acinetobacter Pseudomonas Aeromonas*
Schöbel	1983, 1985 1986	diEG & triEG tetraEG & PEG600	*Pseudomonas fluorescens Alcaligenes glycovorans*
Steber & Wierich	1985	PEG 400	Activated sludge
Obradors & Aguilar	1991	PEG up to 10 000	*Pseudomonas stutzeri*

[a] The growth might be caused by lower molecular PEG included in a sample.
[b] A representative mixed culture was composed of *Sphingomonas terrae* and *Rhizobium* sp.

S. terrae. The associated bacteria in symbiotic mixed cultures are of the genera *Rhizobium*, *Agrobacterium* and *Methylobacterium*. Recently, we isolated individual cultures which grow on PEGs up to 20 000 (unpublished data). It is not known whether such isolates belong to the same *Sphigomonas* species. Even if pure cultures can degrade PEGs, symbiosis is still an important and ubiquitous phenomenon in natural environments, which seems to be involved in biodegradation of recalcitrant materials. The mechanism for symbiotic biodegradation of PEG was elucidated [84] as being due to removal of a toxic metabolite, glyoxylic acid, formed during the biodegradation process, as shown in Fig. 4.

Since an early study by Mills and Stack [85], no reports were presented on anaerobic degradation of PEGs for approximately thirty years. In 1983, three groups reported the anaerobic growth of isolates or consortia on PEGs [86–88]. *Alcaligenes faecalis* var. *denitrificans* was first isolated by Fincher and Payne as an aerobic PEG 400-utilizing bacterium [82]. This strain also grew anaerobically as a denitrifier at the expense of several free ether glycols up to PEG

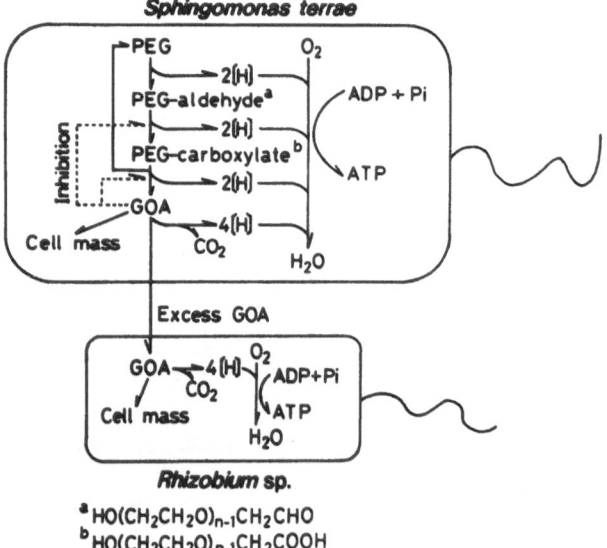

Fig. 4. Symbiotic mechanism of PEG 20 000-utilizing mixed culture E-1. Modified from Fig. 10 of Ref. [81]

300 [88]. Mills and Stack [85] and Taylor et al. [89] reported anaerobic disposal of xenobiotic compounds under nitrate-consuming conditions. Dwyer and Tiedje [87] isolated methanogenic consortia from sewage sludge, which can degrade ethylene glycol monomer and polymers as large as PEG 20 000. Schink et al. isolated anaerobes which can degrade PEG up to the 20 000 size [86].

Poly(1,2-propylene oxide) (PPG) is divided into two groups (straight and branched chain structures) according to its starting compounds. When a chain of 1,2-propane diol is expanded from ethyleneglycol, a straight chain is obtained (diol type); and when glycerol is used in place of ethylene glycol, a branched polymer chain is synthesized (triol type). PPG having an average molecular weight lower than approximately 700 (triol type) or 1000 (diol type) is readily water-soluble, but one having a higher Mn is oily and insoluble in water. Fincher and Payne observed that a PEG-utilizing isolate could assimilate 1,2-propylene glycol and its dimer as sole carbon and energy source [82]. Meanwhile, our PEG-utilizing isolates [81], and those isolated by Watson and Jones [90], did not grow on dimer or PPG. Anaerobic PEG-utilizing bacteria isolated by Schink and Stieb [172] and Dwyer and Tiedje [175] did not degrade PPG either.

PPG-utilizing bacteria were isolated from soils by an enrichment culture containing PPG 2000 or 4000 or from activated sludges acclimated to PPG 2000 or 4000 for a few months [91]. A typical PPG-utilizing bacterium was identified as *Corynebacterium* sp. The organism grew on monomer up to PPG 4000 (triol type) and 3000 (diol type), but did not assimilate PEGs. The bacterium utilized all structural and optical isomers included in the dimer and also isotactic and atactic structures. Generally speaking, stereospecificity of microorganisms and their enzymes is strict. However, in this case, the bacterium either has several stereospecific enzymes for optical isomers or a nonstereospecific enzyme. *Corynebacterium* sp. No. 7 entrapped in polyacrylamide gels could also degrade approximately 80% of 0.1% PPG overnight [81]. Degradation depended on the concentration of acrylamide used for making the gels, i.e., loose gel matrices seemed to bring about better degradation by the easy passage of PPG molecules. Adsorption of oily PPG on the gel surfaces was feared but not detected. The immobilized cells degraded PPG 670 to 2000 (diol type) and 1000 to 4000 (triol type). PPGs having higher Mn were degraded to a greater extent than those having lower Mn. PPG was also removed by a continuous culture of immobilized cells.

The effects of side or main chain structures on the growth of PPG-utilizing *Corynebacterium* sp. No. 7 were examined, as shown in Table 4 [92]. The bacterium grew well on poly(1,2-butane oxide) 400 and 2000, which are oily materials used as lubricants. A methyl group in PPG was replaceable by an ethyl group in PBG. The microorganism grew on polyglycerins to some extent, but not on polyglycidols, PEG, PTMG or polyvinyl alcohol. These results indicated that the bacterium recognized an ether oxygen adjacent to two or three carbon chains and a hydrophobic side group such as a methyl or ethyl group.

PPG was not degraded by a culture filtrate or a cell-free extract of *Corynebacterium* sp., but was metabolized by washed, resting cells. These results suggested that PPG was metabolizable neither by extracellular enzymes nor by a hydrolase, but possibly by intracellular enzymes including membrane-bound enzymes linked to a respiratory chain [93].

PTMG is used exclusively for synthesizing polyurethanes. As water-soluble oligomers are unsuitable for synthesizing polyurethanes, they are removed with water from a mixture of polymers and thus are contained in the waste water of

Table 4. Growth of *Corynebacterium* sp. No. 7 on polymers

Substrate (0.5%)	Growth (OD at 610 mn)	pH
None	0.12	7.0
PEG 400	–	7.0
PPG 2000	1.34	5.8
PBG 400	1.27	5.0
2000	0.92	7.0
CoEO/glycidol 5000 (R, S)	0.22	7.0
Diglycerin	0.66	4.6
Polyglycerin 310	0.49	4.8
500	0.32	5.6
750	0.33	5.8
Polyglycidol 13 300 (R)	–	7.0
13 300 (S)	–	7.0
PTMG 200	–	7.0
PVA n=500	–	7.0

Reprinted from Ref. [92].

a synthetic chemical plant. The water-soluble fractions were analyzed by HPLC and LC-MS and eight peaks were detected. Oligomers up to about a polymerization degree of 8 seem to be water-soluble. PTMG-utilizing bacteria were isolated in the author's laboratory from soil and activated sludge samples by enrichment culture techniques [94]. Although oxidation of 1,4-butanediol by strains of *Gluconobacter* and *Acetobacter* and bacterial utilization of 1,4-butanediol as sole carbon source were reported, no other report on microbial utilization of PTMG has appeared. Two strains that showed high growth on a PTMG medium were selected and identified as *Alcaligenes denitrificans* subsp. *denitrificans* and *Xanthomonas maltophilia*. PTMG 265 or 200, which includes monomer to octamer, completely disappeared in 7 days from the culture filtrates of both strains. PTMG 265 was also metabolized by intact cells. The oligomers with a higher polymerization degree disappeared more rapidly than those with a lower polymerization degree. Sonic extracts of *A. denitrificans* dehydrogenated PTMG when coupled with an artificial electron acceptor (DCIP, MIT or ferricyanide) and PMS. These results suggest that PTMG is degraded via an oxidation process linked to an electron transport system of the bacterium. During the bacterial oxidation of PEG and PPG, the oxidation of terminal alcoholic groups takes precedence over the cleavage of an ether bond, as described above. PTMG might possibly be degraded by the same mechanism as in the degradation of PEG and PPG.

In conclusion, polyethers used as water-soluble or water-miscible specialty polymers are susceptible to biodegradation, although biodegradation rates are dependent on the chemical structure, Mn and physical properties.

3.4 Polyamides

There are many reports on the enzymatic degradation of synthetic poly-L-amino acids. Although the peptide linkage in proteins (random polymers of various amino acids) is similar to that of the amide linkage in synthetic polyamides (homopolymers of single amino acids), the observed degradation rates of proteins are generally much faster than those of synthetic polyamides. Peptidases seem to be stereospecific for peptides and polyamides. Huang et al. discussed the effect of structure and morphology on the degradation of polymers with multiple groups, including polyamides [71]. Conformational flexibility leads to biodegradability and morphology of the polymers affects the rate of degradation.

Ebata et al. reported on the enzymatic degradation of 6-nylon oligomers by trypsin [95]. Linear and cyclic 6-aminocaproic oligomer-utilizing bacteria have been isolated, which can utilize oligomers (polymerization degrees= 1–6) as sole carbon or carbon and nitrogen sources [96–98]. The oligomers were hydrolyzed by cyclic dimer hydrolase, and exogenous and endogenous 6-aminocaproic oligomer hydrolases [99]. These enzymes are coded on plasmid pOAD2 of *Flavobacterium* sp. K172 which assimilates 6-nylon oligomers [100]; total DNA of the plasmid has been sequenced [101, 102]. However, no reports on higher oligomers and polymers have been published.

3.5 Polyurethanes

Polyurethane (PU) has a wide range of applications such as in furniture, bedding, automobiles, constructional materials, elastmers, adhesives, paints, fibres and synthetic skin. The polymer is commonly prepared by reacting isocyanates and polyols. In terms of polyols, petrochemicals such as polyalkylene glycol and polyesterpolyol are used. According to the polyols used, there are two types of PUs; polyester PU and polyether PU. It is established in general terms that polyester PU is more susceptible to microbial attack than polyether PU [68]. The polyester segment of the PU structures is the initial site of attack. Some PU-deteriorating fungi were capable of inducing esterase/lipase activity in the presence of poly(caprolactone) diol [103]. Polyester PU was hydrolyzed by a commercial lipase [77]. The biodeterioration of PU was well documented by Seal and Pathirana [104].

Natural polymers contain more than two hydroxyl groups in the main chain and can be used as polyols for polyurethane preparation. Several attempts have been made to prepare PU from lignin. Hatakeyame et al. [3] prepared PU from lignocellulose (coffee bean parchment and wood meal from pine) dissolved in PEG 400. By changing PEG content, various types of PU foams were obtained, e.g. from soft type foam to hard type foam, which showed the compressible stress-strain behaviours similar to those of PU foams from petrochemicals. These polymers show excellent biodegradability. Also PU formed from cellulose triacetate and PPG, as well as amylose and polybutadiene/PPG, were reported [105, 106] to be susceptible to cellulase and amylase, respectively.

As for enzymatic degradation of the urethane bond, Zhao et al. suggested that ether bonds in PU were nonenzymatically hydrolyzed but papain hydrolyzed urethane and urea bonds included in PU [107]. It is an interesting point whether the enzyme is similar to β-ureidopropionase (E.C. 3.5.1.6), which can cleave an ureido bond.

3.6 Copolymers

The purposes of copolymers include better formulation and better biodegradability than the component polymers. Knowledge of the biodegradability of various polymer structures suggests that oxidizable and hydrolyzable chemical structures are susceptible to biodegradation. Therefore, biodegradable copolymers have to include oxidizable and hydrolyzable groups as random copolymers or block copolymers: ester, ether, amide and PVA. Other biodegradable copolymers are a combination of natural polymers, which are inherently biodegradable, and synthetic polymers which should be biodegradable.

As sodium tripolyphosphate (STPP) shows excellent builder performance, but causes a serious eutrophication problem, various substitutes have been investigated. Poly(sodium carboxylate)s, such as poly(sodium acrylate), give excellent builder performance when compared with STPP on an equal weight basis. These compounds, however, are extremely resistant to biodegradation. Matsumura and Yoshikawa synthesized various water-soluble polymeric polycarboxylates containing carbon-carbon backbones and a high charge density of carboxylate groups along the chain [108]. They compared their biodegradability and builder performance in detergents. Introduction of an ester to the backbone and partially oxidized polysaccharides containing unreacted glycopyranosyl groups were the most effective as biodegrading segments in the polymer. Polyvinyl-type poly(sodium carboxylate)s and poly(sodium carboxylate)s containing glycopyranosyl groups showed better builder performance in detergents, compared to those containing ester or ether linkages.

Yamamoto et al. [109] and Yasuda et al. [110] synthesized aliphatic copolyesterethers from cyclic esters; β-propiolactone, β- and γ-butyrolactone, β-methyl-δ-valerolactone, ε-caprolactone (CL), glycolide, lactide, and cyclic ethers like ethylene oxide (EO), propylene oxide (PO) and isobutylene oxide. Biodegradability of these copolyesterethers were tested with two lipases (*Candida cylindracea* and *Rhizopus arrhizus*) and hog liver esterase [109], with which all copolymers were hydrolyzed. The introduction of a methyl group tended to decrease degradability; the hydrophilic EO-copolymers were more rapidly degraded, compared with the hydrophobic corresponding PO-copolymers; the copolymers of PO with GA and LA of high melting point or crystallinity suppressed degradability. Copoly(EO/ester)s (EO: 15–90%) were assimilated by the PEG-utilizing symbiotic mixed culture as sole carbon source [111], although the growth depended on the EO content of a copolymer. On the other hand, copoly(PO/CL)s (PO: 24–52%) were not utilized by PPG-utilizing *Corynebacterium* sp. No. 7 as sole carbon source. A terminal hydroxyl group of the PPG segment might be blocked or any physical property might be inhibitory for microbial attack.

PEG-PPG block copolymers were assimilated by PEG- and/or PPG-utilizing bacteria [94], as shown in Table 5. Block polymers *a* and *d* which included a larger amount of PPG were utilized by *Corynebacterium* sp. No. 7 as sole carbon source. This suggested that the terminal structure is possibly PPG. Polymers *c* and *f*, were not attacked by PPG-utilizing bacteria, but they were attacked by symbiotic mixed culture E-1 as sole carbon source. Polymers *b* and *e* were not assimilated by either culture, possibly because of their toxicity to living cells at 1.0%. However, since PEG dehydrogenase oxidized them, the compounds are thought to be metabolizable.

Microbial P(3HB) has attracted industrial attention as an environmentally degradable thermoplastic for a wide range of agricultural, marine, and medical applications. However, it is a highly crystalline and has brittle characteristics. To improve the physical properties of the compound, blending with synthetic polymers, polysaccharides, and microbial PHA has been attempted together with the production of a variety of PHAs which have improved physical properties. It has been reported that PEG is miscible with P[3HB] in the amorphous state [112]. In addition, a block copolymer of atactic P[(R,S)-3HB] and PEG segments which has an Mn of 10 500 and A-B-A triblock structure was further blended with microbial P(3HB) [113]. The P(3HB) film became flexible and tough by

Table 5. Microbial utilization or enzymatic degradation of PEG-PPG block copolymers

Copolymer	Growth (E_{610}) (1)	(2)	PEG Dehydrogenase (10 mM)*
a	1.3	0.6	
b	0.3	0.4	54
c	0.3	2.1	
d	1.1	0.8	
e	0.2	0.3	145
f	0.3	1.4	

* The enzyme activity of *S. macrogoltabidus* No. 203 toward PEG 4000 was defined as 100.
Strains used: (1) *Corynebacterium* sp. No. 7; copolymer, 0.5%
(2) Symbiotic mixed culture E-1; copolymer, 1.0%
Copolymers: PEG-PPG-PEG, Dai–Ichi Kogyo Seiyaku Co., Ltd., Kyoto, Japan.

Copolymer	MW	PPG (MW)	PEG (MW)
a	1330	1200	130
b	2400	1200	1200
c	8000	1200	6800
d	2220	2000	220
e	4000	2000	2000
f	13 000	2000	11 330

Reprinted from Ref. [111].

means of blending with the block copolymer. Enzymatic degradation of the film by an extracellular PHB depolymerase from *A. faecalis* took place solely on the surface of the blended films.

Polyesters PGA, PLA, PMA and PHB are mainly used as medical polymers. These materials are also used as copolymers such as PGA-PLA, PMA-PPG, PGA-PEG and PLA-PEG. Cohn [114] reported the synthesis of tailor-made polymers by copolymerizing PGA and PLA with polyethers, in which partially crystalline polyesters create the hard domains, while the elastic response of the system is mainly provided by the flexible polyethers (PEG, PPG and PTMG). Polyethers also contribute hydrophilicity vs. hydrophobicity. Albertsson et al. synthesized a block polymer of polyethylene succinate and PTMG/PEG as an elastic absorbable suture which could be hydrolyzed in vivo [115].

Starch- and vinyl alcohol copolymers have been commercialized under the Mater-Bi trademark [116]. Starch, even if significantly shielded by the interpenetrated structure, is first hydrolyzed by extracellular enzymes; the synthetic component is biodegraded at a slower rate by microorganisms adsorbed on the plastic surface. Biodegradation is made easier by the increase in available surface during the hydrolysis of the natural component.

Baily et al. synthesized polyesterolefins which were susceptible to biodegradation by soil microorganisms [117]. Huang et al. produced poly(amide-esters) which were hydrolyzed by subtilisin [71].

3.7 Polymer Blends

Polymer blends, particularly olefins with biodegradable polymers, are gaining popularity as an approach to degradable packaging plastics. The materials are at best only partially biodegraded, but will lose form and bulk as the plastic disintegrates. This may be sufficient in landfill as volume diminishes, leaving room for more solid waste. It, however, should be understood that only the biodegradable part of the blend has been shown to be removed from the environment.

This work was started by Otey [118–120], who developed starch-based polyethylene films, and Griffin [121–123] who introduced the concept of using corn starch as a filler to accelerate the degradation process (Fig. 5) [189]. Wool and Cole [124] investigated the degradability of corn starch-based polyethylenes and modeled the biodegradation by a percolation theory. When the plastic blend is placed in a biologically-active environment, starch is thought to accelerate the degradation rate by microbial attack on the polysaccharide. The microbes subsequently invade the plastic by consuming starch and creating pores in the process. This provides greater surface contact between the microbes and the PE, speeding up biotic reactions on the synthetic plastic. Oxygen may then adsorb on the newly created surface area, enhancing abiotic reactions. Macroorganisms such as crickets, roaches, slugs and larger animals can also contribute to the degradation through consumption and ingestion of these blends. Photodegradation via Norrish–Smith type reactions may occur through UV absorbing groups such as ketones in the PE matrix [125].

Starch is recognized as a cheap filler for polyolefins, enhancing their biodegradability, or apparent biodegradability, depending on the interpretation of results: starch/poly(vinyl alcohol) [126], starch/polyurethane [127], and starch/poly(methyl methacrylate) [128] also biodegrade. Potts has patents in this area with olefin/polycaprolactone blends [129]. Starch/P(3HB-co-3HV) films show superior biodegradability and practical-level thermoplasticity [16]. Other uses of blends include controlled rate of fertilizer release [130] based on ethylene/vinyl acetate/carbon monoxide polymers which are UV-sensitive, polyolefin blends with any biodegradable polymer [131] and polyolefins blended with metals and autoxidizable substrates [132]. Starch, being cheap, will continue to be an attractive substrate [133] unless the blends and grafts are found to be unacceptable substitutes for packaging materials.

4 Enzymes Involved in the Metabolism of Polymers

4.1 Polyesters and Polyhydroxyalkanoates

In nature, polyesters exist as cutin (in plants) and as PHB (in bacteria). Polyesters are known to be non-enzymatically hydrolyzed by water, but enzymatic hydrolysis accelerates degradation. Potts established very early that only low melting and

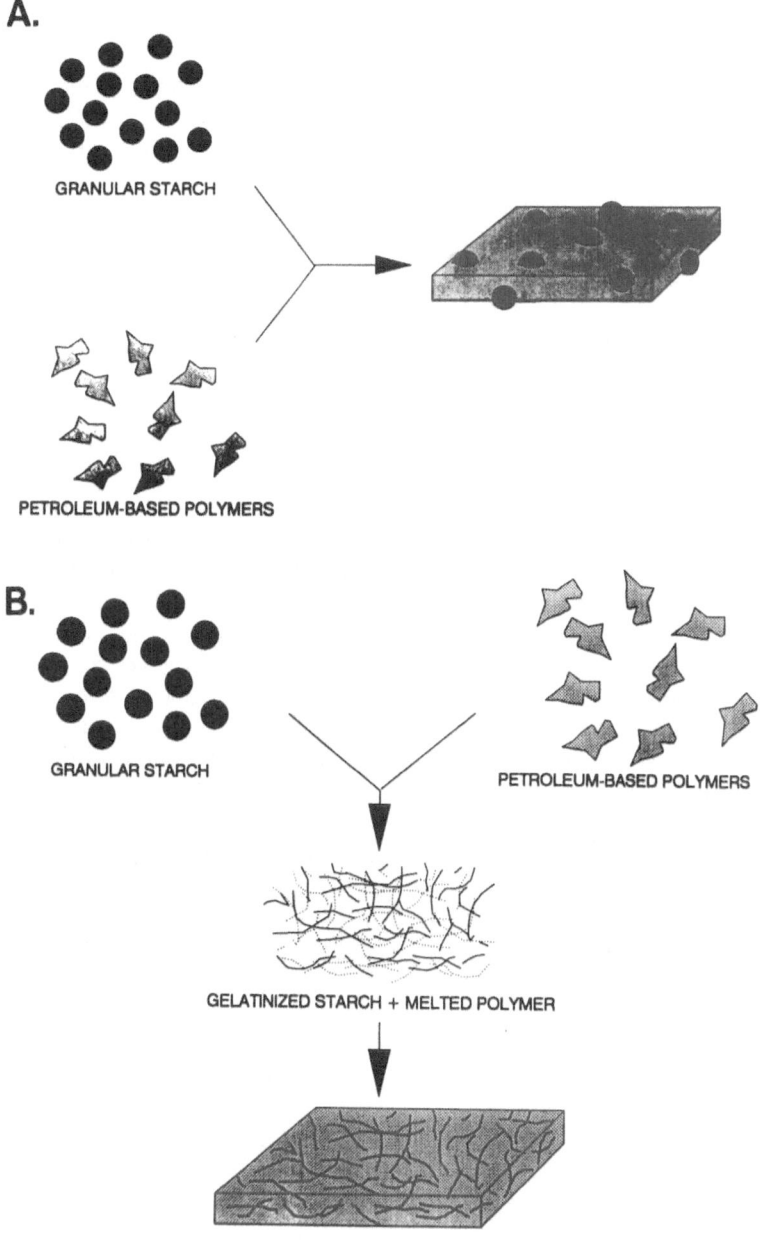

Fig. 5. Methods for producing plastic composites containing granular and gelatinized starch. A: Griffin's Method, B: Otey's Method. Reprinted from Ref. [189]

low molecular weight aliphatic polyesters were biodegradable [134]. Kendrick demonstrated that amorphous region of polyesters were more readily biodegraded than crystalline regions [135].

Tokiwa et al. isolated a polyethylene adipate-utilizing fungus, *Penicillium* sp. 14–3 and a polycaprolactone-utilizing fungus, *Penicillium* sp. 26–3 [72, 73]. PEA and PCL were hydrolyzed by extracellular enzymes, respectively. They purified an extracellular enzyme from *Penicillium* sp. 14–3 to homogeneity on disc electrophoresis [74]. The purified enzyme had a wide substrate specificity for saturated and unsaturated aliphatic polyesters, PCL, alicyclic polyesters, plant oils, triglycerides and fatty acid methyl esters; PHB, and aromatic polyesters were not hydrolyzed. These results suggest that the enzyme is a lipase. They also studied the enzymatic degradation of polyesters by various commercially available lipases and esterases [75–77]. *Rhizopus arrhizus* and *Rhizopus delemar* lipases were especially capable of hydrolyzing various kinds of polyesters [77], as shown in Table 6.

Bioabsorbable polyesters like PGA and PLA are thought to be absorbed in vivo by nonenzymatic hydrolysis and also by enzymatic hydrolysis where various hydrolytic enzymes, e.g., esterases and lipases, seem to be involved.

Cutin is a natural polyester of plants. This compound is degraded by cutinases excreted by pathogenic fungi, which cause infection through the cuticular layer of a host plant. Hog pancreatic lipase also acts on cutin [136]. Thus, polyesters are generally degraded by various lipases and esterases, irrespective of their origins.

PHB is another well-known polyester produced as storage material in various bacterial cells, as described in Sect. 2.2. Chowdhury first reported on P(3HB) depolymerase of *Pseudomonas* which had a wide substrate specificity [137]. Later, P(3HB) depolymerases were detected and purified from various microorganisms: *P. lemoignei* [138], *Alcaligenes faecalis* T1 [139, 140], *Comamonas* [141], *P. picketii* [142], *Comamonas testosteroni* [143], unidentified bacteria [144] and *Penicillium funiculosum* etc. [145]. These enzyme reacted on P(3HB), P(3HP) and P(4HB), but not on other PHA and aliphatic polyesters. On the other hand, poly(3-hydroxyoctanoate), P(3HO), depolymerase was purified from *P. fluorescens*, which assimilated P(3HO) [146]. The enzyme hydrolyzed copolymers of medium-chain-hydroxyalkanoic acids (C_6–C_{14}) and P(3HO), but P(3HB) and P(3HV) could not be hydrolyzed. These enzymes, referred to as PHA depolymerases, were excreted in culture supernatants, which corresponded well to the water-insolubility and high-molecular weights of PHA and PHB. P(3HB) depolymerase of *A. faecalis* T1 is well characterized: it has P(3HB) depolymerizing and 3HB oligomer hydrolyzing sites and is thought to be an endo-type esterase [140, 147]. The reaction products of the former enzyme are 3HB oligomers, which are further hydrolyzed into monomers via lower oligomers. Commercial lipases hydrolyzed PHA except P(3HB) [148]. Thus, PHA depolymerases are divided into three groups according to their substrate specificities toward PHAs, the principles of which are quite similar to each other and might have evolved from lipases.

Table 6. Hydrolysis of Polyesters by *R. delemar* and *R. arrhizus* Lipases

Polyester	$\overline{M}n$	Tm (°C)	Powder size	TOC formed by lipases (ppm) R. delemar	R. arrhizus
Polyethylene adipate	2720	48.5	C	8360	9290
Polyethylene suberate	4050	64.5	B	1020	1620
Polyethylene azelate	4510	52.1	B	3080	3770
Polyethylene sebacate	1570	74.5	A	550	980
Polyethylene decamethylate	1610	86.0	B	180	240
Polytetramethylene succinate	4240	117	A	150	210
Polytetramethylene adipate	1790	72.0	B	3360	2900
Polytetramethylene sebacate	2440	65.8	A	980	3300
Polyhexamethylene sebacate	5820	74.0	B	380	1160
Poly-2,2-dimethyltrimethylene succinate	2370	76.5	A	240	50
Poly-2,2-dimethyltrimethylene adipate	2020	36.5	E		150
Polyglycolide		226–234	E		0
Copolyester of glycolide and lactide (molar ratio, 92:8)		200–210	E		0
Polypropiolactone	4270	95.0	A	2240	1600
Poly-DL-β-methylpropiolactone	8190	167–171	E	10	60
Poly-D -β-methylpropiolactone (PHB)	25000	175–181	A	0	0
Polycaprolactone (PCL)	6740	59.0	A	310	3610
Poly-cis-2-butene adipate	2700	56.9–59.8	C	580	550
Poly-cis-2-butene sebacate	6190	60.8–62.5	B	300	3430
Poly-trans-2-butene sebacate	3560	57.0–59.0	A	340	1190
Poly-2-butyne sebacate	4930	61.9–63.0	C	670	910
Polyhexamethylene fumarate		113–117	A	35	0
Poly-cis-2-butene fumarate		300	B	30	40
Polytetramethylcyclobutane succinate	3440	63.5–84.0	E	0	0
Polycyclohexylenedimethyl succinate	3910	123–130	B	130	120
Polycyclohexylenedimethyl adipate	3250	108–114	B	200	160
Polytetramethylene terephthalate		230–240	C	0	0
Polyethylene tetrachlorophthalate	1670	78.0–84.2	A	0	0
Poly-2,2-dimethyltrimethylene isophthalate		66.5–75.0	A	0	0
Poly-p-hydroxybenzoate		300	A		0
Poly-3,5-dimethyl-p-hydroxy-benzoate		300	A		0
Poly-4-ethoxy-3,5-dimethyl-benzoate		259–267	A		0

The particle size of each polyester powder was ranked A, B, C, D, or E, corresponding respectively to roughly less than 0.25 mm, less than 0.50 mm, less than 1.0 mm, 0.25–1.5 mm, 0.25–3 mm. Each reaction mixture contained 400 µmol of phosphate buffer (pH 7.0), 1 mg of surfactant Plysurf A210G, 300 mg of the polyester powder and 60 µg of *R. arrhizus* lipase (or 300 µg of *R. delemar* lipase) in a total volume of 10.0 ml. In the case of *R. delemar* lipase, surfactant was omitted and pH of phosphate buffer was 6.0. In the substrate and enzyme controls, enzyme or substrate was omitted from the reaction mixture. The reaction mixtures were incubated at 30 °C for 16 hours.
Reprinted from Ref. [77].

Genes for PHA depolymerases have been cloned: *A. faecalis* T1 [149], *P. lemoignei* [150] and *P. picketti* K1 [151]. A gene for P(3HB) depolymerase *c* of *P. lemoignei* and genes from two other bacteria had similar sequences, which suggested the evolution of enzymes from the same origin.

Only recently, reductive and methanogenic degradation of PHB was reported: This could be an efficient means of dissolving PHB in anoxic environments such as sanitary landfills [152, 153].

4.2 Polyethers

Biodegradation rates of polyethers and their derivatives were assayed with activated sludge from municipal sewage plants without any further acclimation, as shown in Table 7 [92]. High biodegradation rates with PEG, PPG and PTMG coincided with the isolation of PEG, PPG or PTMG-utilizing bacteria from soils or acclimated activated sludges. PGB also showed high biodegradability. As described in Sect. 3.3, this compound was assimilated as sole carbon source by a PPG-utilizing bacterium. Thus, polyethers are biodegradable in general. However, dialkyl derivatives of PEG showed very poor biodegradabilities. This suggested that at least one free alcohol group is necessary for biodegradation.

Table 7. Biodegradation of polyethers and their derivatives by activated sludges[a]

Substrate		Mn	Biodegradation (%)[b]	
			7-day	14-day
PEG	$HO(EO)_n H$	390	99.4	99.7
		1500	83.9	95.9
PEG-dimethyl	$MeO(EO)_n ME$	1500	7.4	9.4
PEG-diethyl	$EtO(EO)_m CH_2 O(EO)_n Et$	1600	5.5	1.3
PEG-dibutyl	$BuO(EO)_m CH_2 O(EO)_n Bu$	1500	11.0	10.8
PPG	$HO(PO)_n H$	410	60.1	69.7
PBG	$HO(BO)_n H$	330	58.3	82.5
PTMG	$HO[(CH_2)_4 O]_n H$	660	31.0	99.8
CoEO/PO[c]	$HO(EO/PO = 75/25)_n H$	1500	48.4	66.6
	$HO(EO/PO = 50/50)$	1470	33.4	49.0
	$HO(EO/PO = 25/75)$	1390	49.7	57.4
	$HO(EO/PO = 75/25)$	12000	15.6	16.4
CoEO/ester	$MeO(EO)_m CO(CH_2)_4 COO(EO)_n Me$			
		1700	96.5	99.4
CoEO/urethane	$HO(EO)_m CONH(CH_2)_6 NHCOO(EO)_n H$			
		20000	10.6	37.3

EO: ethylene oxide, PO: 1,2-propylene oxide, BO: 1,2-butylene oxide, Me: methyl,Et: ethyl and Bu: butyl.
[a] From municipal sewage plants.
[b] Measure by total organic carbon.
[c] Random polymerization.
Reprinted from Ref. [92].

Copoly(EO/PO) with an average molecular weight of approximately 1500 were degraded well, irrespective of EO/PO ratios, but a copolymer with a higher Mn showed decreased biodegradability. The high biodegradability of dimethyl PEG, which includes an internal ester bond, can be explained by the easy hydrolysis of an ester bond: the ester bond appeared to be hydrolyzed first and then the resultant alcohol groups of PEG seemed to be metabolized. A PEG derivative which included a urethane bond inside a PEG molecule showed reduced biodegradability, but an approximate 40% biodegradation rate was observed in 14 days. An urethane bond might be hydrolyzed to PEG or the compound might be metabolized from a terminal alcohol group.

Dehydrogenations of PEGs linked to electron acceptors such as NAD, flavins and ferricyanide were observed with enzyme preparations obtained from PEG-grown bacteria [81]. Oxidation of terminal alcohol groups was also suggested by the work of Kawai [81] and Watson and Jones [90]. Harada and Nagashima [154] isolated metabolites from a 6-day culture of *Alcaligenes* sp. incubated with EG monoethyl ether or monomethyl ether and identified ethoxy and methoxy acetic acid. Patterson et al. [155] detected acidic metabolic intermediates from degradation of alcohol ethoxylate. The formation of mono- and diacid metabolites of the glycol in mammals has also been reported [156]. Thus, oxidation of terminal alcoholic groups to carboxyl groups appears to be the common pathway in PEG metabolism.

Hosoya et al. [157] obtained dicarboxylic acids of ethylene glycol (EG) dimer and trimer from a culture filtrate of bacteria grown on the trimer and suggested that PEG might be degraded oxidatively by splitting a C_2 unit from a terminal alcohol which is first oxidized to a carboxyl. Author et al. detected mono- and dicarboxylic acids of EG tetramer (TEG), and depolymerized oligomers (trimer, dimer and monomer) in a reaction mixture with TEG [158]. In addition, TEG-aldehyde was characterized as a reaction product of TEG incubated with a crude PEG dehydrogenase (cell-free extracts) [159]. Valeraldehyde was also formed by cell-free extracts from *n*-pentanol. Thus, the metabolic pathway for PEG was concluded to occur as shown in Fig. 6. PEG is successively oxidized to an aldehyde and a monocarboxylic acid, which is followed by the cleavage of the ether bond, resulting in PEG molecules that are shortened by one glycol unit. Simultaneous oxidation of two terminal alcohol groups of the molecule is also possible. Depolymerization might proceed via the same reaction observed with the monocarboxylic acid. This sequence is repeated and eventually yields depolymerized PEG. The resultant glyoxylic acid may then enter into central metabolic routes by known pathways: e.g., the oxidative dicarbonic acid cycle, tricarbonic acid cycle, and the glycerate pathway. The degradability of PEG with different terminal structures was examined (Table 8) [92]. PEG-diglycolic acid, a metabolite of PEG, was utilized by the PEG 20 000-utilizing symbiotic mixed culture E-1 (*Sphingomonas terrae* and *Rhizobium* sp.). Monoalkyl PEG was utilized, but dialkyl PEG was not, suggesting that degradation of PEG is started from a terminal alcohol group. This inference was supported by the observation that depolymerized products were not detected during the degradation of PEG by the

OXIDATIVE METABOLIC PATHWAY OF PEG

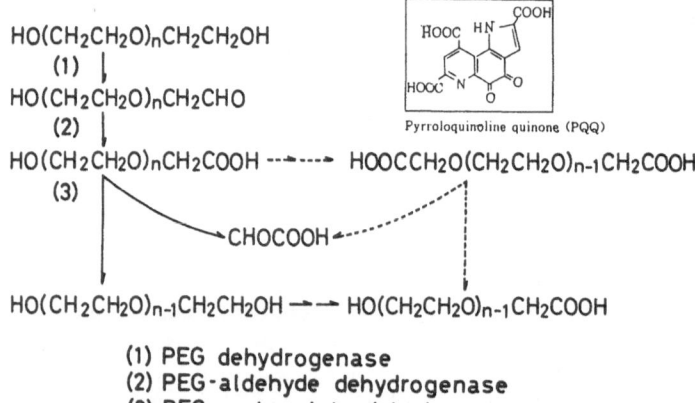

$$HO(CH_2CH_2O)_nCH_2CH_2OH$$
$$(1) \downarrow$$
$$HO(CH_2CH_2O)_nCH_2CHO$$
$$(2) \downarrow$$

Pyrroloquinoline quinone (PQQ)

$$HO(CH_2CH_2O)_nCH_2COOH \dashrightarrow HOOCCH_2O(CH_2CH_2O)_{n-1}CH_2COOH$$
$$(3)$$

$$CHOCOOH$$

$$HO(CH_2CH_2O)_{n-1}CH_2CH_2OH \longrightarrow HO(CH_2CH_2O)_{n-1}CH_2COOH$$

(1) PEG dehydrogenase
(2) PEG-aldehyde dehydrogenase
(3) PEG-carboxylate dehydrogenase

Fig. 6. Aerobic metabolic pathway of PEG

Table 8. Growth of the mixed culture E-1 on PEG derivatives.

Substrate (0.5%)	Growth (OD at 610 nm)[a]
PEG monomethyl ether Mn: 350	1.74
550	2.36
750	2.74
1900	1.90
5000	1.67
Diethylene glycol monoethyl ether	0.78
Diethylene glycol diethyl ether	–
Triethylene glycol monomethyl	0.94
Triethylene glycol monobutyl	1.32
Triethylene glycol monochlorohydrin	–
Triethylene glycol dimethyl ether	–
Tetraethylene glycol dimethyl ether	–
PEG diglycolic acid Mn; 400	1.03
1000	0.96
3000	0.56
PEG-distearate	–
PEG-laurylamine	–
PEG-lauryl ether (Brij 35)	–
PEG-cetyl ether (Brij 58)	0.97
PEG-tetrahydrofurfuryl ether	0.68
PPG	–
PTMG	–

[a] 5–10 days at 28 °C with shaking.
–: No growth.
Reprinted from Ref. [92].

mixed culture. In other words, the endogenous breakdown of a polymer molecule rapidly yields depolymerized products to some extent. On the other hand, chemical degradation of PEG by Fenton's reagent suggests endogenous breakdown of a molecule from a gel-permeation-chromatography (GPC) pattern of degradation products [139]. The culture neither grows on PPG nor PTMG. Thus, biodegradation proceeds exogenously from a terminal group and strictly depends on the chemical structures of the monomer units.

Haines and Alexander [160] reported that PEG 20 000 might be hydrolyzed by an extracellular enzyme of *Pseudomonas aeruginosa* to yield oligomers as metabolic products, but these results have not be reproduced and the original strain was lost.

Pearce and Heydeman [161] suggested a non-oxidative removal of EG units as acetaldehyde by a membrane-bound, novel oxygen-sensitive enzyme, diEG lyase. Of cofactors tested, only cyanocobalamine and adenosylcobalamine stimulated the reaction, but this varied from one preparation to another. They assayed enzyme activity by vapor-phase chromatography of the incubation mixture. However, the measurement of compounds in the presence of fractionated cell protein material is risky at best due to the presence of unidentified metabolizable materials.

Schöberl [162] first suggested that PEG was catabolized by a C_1 step, liberating formate which is metabolized by a serine pathway. Then he [163] corrected the C_1 hypothesis and reported that dicarboxy products were obtained from dimer $\sim$ tetramer. He suggested that an ether bond was hydrolyzed to liberate glycolic acid.

Thélu et al. [164] reported that cell-free extracts from *Pseudomonas* sp. grown on PEG 400 dehydrogenated 2-ethoxyacetic acid or dimethylethyl derivatives of diEG and tetraEG as well as PEG and suggested the transient formation of a double bond in a terminal glycol unit, followed by hydration, according to: $R-O-CH_2CH_2OH \rightarrow [R-O-CH=CHOH] \rightarrow R-CHOHCH_2OH$. Recently, Obradors and Aguiler [165] isolated *Pseudomonas stutzeri* which degraded PEG 10 000 completely and suggested that periplasmic PEG dehydrogenase yielded glyoxylic acid as a product.

We still cannot deny possible routes for PEGs other than the oxidative degradation pathway via carboxylated PEG. However, such other hypothetical routes necessitate further studies.

All the PEG-utilizing bacteria isolated by the author and coworkers showed membrane-bound dehydrogenase activities toward various PEG species [166]. Degradation of PEG by culture supernatants or cell-free extracts was not found [158]. These results suggest that the metabolism of PEG is in the periplasmic space and linked with an oxidative respiratory chain as shown in Fig. 7. These dehydrogenases are induced by PEGs which can be a sole carbon and energy source, respectively, except that the enzyme of PEG 4000-utilizing *Sphingomonas macrogoltabidus* No. 203 is constitutive [166]. PEG dehydrogenases were purified from the symbiotic mixed culture E-1 and from *S. macrogoltabidus* No. 203 [167, 168]. Both enzymes were solubilized with surfactants and stabilized with 10% glycerol or ethylene glycol. The purified enzymes showed quite similar

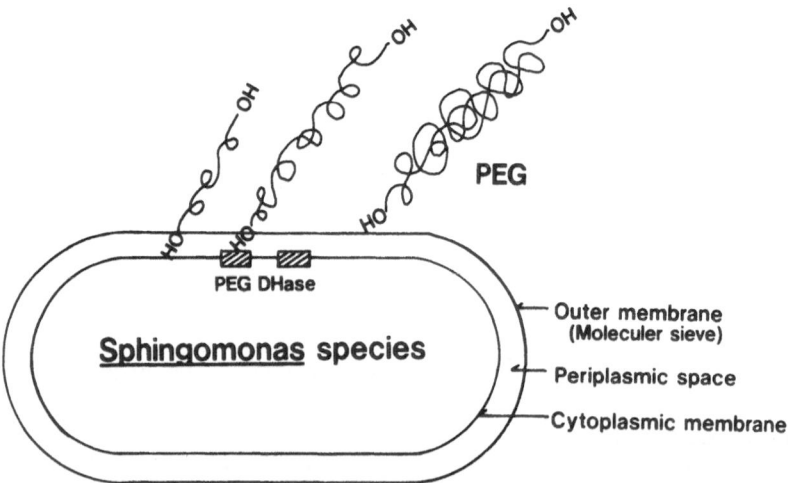

Fig. 7. The suggested metabolism of PEG in the periplasmic space. Reprinted from Ref. [190]

physical and chemical characteristics, as shown in Table 9. The prosthetic group of the enzyme from the symbiotic mixed culture E-1 was isolated and identified as pyrroloquinoline quinone (PQQ) [169]. The enzyme was quite different from other alcohol dehydrogenases reported so far in its substrate specificity and coenzyme. Hence, it is considered to be a novel quinoprotein and long-chain primary alcohol dehydrogenase. The prosthetic group of the enzyme from *S. macrogoltabidus* was also suggested to be PQQ. Purified enzymes oxidized aldehydes appreciably: PEG-aldehyde could be oxidized by PEG dehydrogenases. Yamanaka [170] demonstrated that dye-linked alcohol dehydrogenase of a photosynthetic bacterium, *Rhodopseudomonas acidophila* M402, had high activity toward PEGs up to 20 000, although PEG neither induced nor accelerated PEG dehydrogenase activity. This could be explained by the wide substrate specificity of the alcohol dehydrogenase.

An ether-cleaving enzyme is the recent focus of the biochemistry of PEG metabolism. Since glyoxylic acid was suggested as a reaction product from diglycolic acid (DGA), which corresponds to dicarboxylic acid of the dimer and can be a model for carboxylated PEG, and depolymerized products (trimer, dimer and monomer) were produced from TEG, an ether bond has to be cleaved between the ether oxygen and β-carbon from the end of the polymer chain. We have detected DGA dehydrogenase activity in the particulate fraction of cell-free extracts of a symbiotic mixed culture grown on PEG 6000 [171], but failed to solubilize and characterize the enzyme. For most of PEG-utilizing bacteria, DGA is not a growth substrate. The symbiotic mixed culture showed very low dehydrogenase activity toward the dimer. Therefore, DGA does not seem to be a proper model substrate for the ether-cleaving enzyme. Only recently, we have obtained a more suitable model compound and are preparing for further studies.

Table 9. Comparison of characteristics of PEG dehydrogenases purified from symbiotic mixed culture E-1 (*Sphingomonas terrae*) and *Sphingomonas macrogoltabidus* No. 203

	PEG 6000-utilizing mixed culture E-1	PEG 4000-utilizing bacterium No.203
Localization	membrane particles	membrane particles
Induction	PEG 6000	constitutive
Electron Acceptor	DCIP	DCIP + PMS
Molecular Wt	$6.0–6.2 \times 10^4$ (SDS-electrophoresis); 2.4×10^5 (Sephadex G-200)	$5.7–5.8 \times 10^4$ (SDS-electrophoresis); 2.2×10^5 (Toyopearl HW-55)
Substrate Specificity	primary alcohols ($C_2–C_{16}$), 1,4-butanediol, PEGs (dimer-20,000)	primary alcohols ($C_2–C_{12}$); diols ($C_3–C_8$); PEGs (dimer-20 000)
Kms	PEG 6000: 3 mM Tetramer: 10 mM	PEG 6000: 5.9 mM PEG 4000: 2.8 mM PEG 1000: 1.7 mM PEG 400: 1.0 mM
Stability	pH 7.5–9.0, 0–35 °C	pH 6.0–8.0, 0–40 °C
Optimum	pH 8.0–8.5, increased up to 60 °C	pH 7.0–8.0, increased up to 35 °C
Effect of Reagents and Metal Ions	inhibited by SH reagents and heavy metal ions; no effect by chelating agents and 2-mercaptoethanol; strongly inhibited by 1,4-benzoquinone (2×10^{-5} M)	inhibited by p-CMB and heavy metal ions; no effect by chelating agents; strongly inhibited by 1,4-benzoquinone (2×10^{-5} M)
Prosthetic Group	PQQ	PQQ

PPG-utilizing *Corynebacterium* sp. No. 7 grew on monoalkyl PG oligomers, but not on PG monoalkyl acetate. Therefore, free terminal alcohol groups were necessary for biodegradation of PPG. The aerobic metabolism of PPG was studied using dipropylene glycol (DPG) as a model substrate for biodegradation [93]. By GC-MS analysis, structural and optical isomers were separated and determined, as shown in Fig. 8. The ratio of structural isomers A, B, and C was 36.3, 48.5, and 15.2%, respectively. With appropriate shaking, metabolic products were accumulated in the culture filtrate. Isomer A was degraded by 51.4%. The diastereomers of isomer B were equally degraded by 40.7% and those of the isomer C were also equally degraded by 18.5%. These results suggest that the bacterium has a wide substrate specificity not only toward structural isomers where secondary alcohol groups were preferentially oxidized, but also toward optical isomers. Assimilation of PPG which is a random copolymer of R,S suggested low selectivity towards R and S. Furthermore, PPG was neither degraded by a culture filtrate nor by a cell-free extract. Instead, intact cells or cell debris, which includes unbroken cells, degraded PPG. Hence, degradation of PPG seemed to be

Structural   Optical   **Fig. 8.** Structural and optical isomers included in
Isomer   Isomer   PG dimer

A $\overset{CH_3}{HOCHCH_2}\text{-}O\text{-}\overset{CH_3}{CH_2CHOH}$ R,R
                                          S,S
                                          meso

B $\overset{CH_3}{HOCHCH_2}\text{-}O\text{-}\overset{CH_3}{CHCH_2OH}$ R,R
                                          S,S
                                          R,S
                                          S,R

C $HOCH_2\overset{CH_3}{CH}\text{-}O\text{-}\overset{CH_3}{CHCH_2OH}$ R,R
                                          S,S
                                          meso

* asymmetric carbon

catalyzed by intracellular enzymes including membrane enzymes through several steps, not by a simple hydrolysis. PPG 2000 and DPG were oxidized by the cell-free extract linked with the DCIP-PMS system. From the results described above, it seems likely that PPG might be metabolized and depolymerized via the same mechanism as that for PEG: oxidation of terminal alcohol groups leading to the cleavage of the ether linkage. PG was also produced from the dimer suggesting that an ether cleavage occurred between the ether oxygen and the adjacent β-carbon of oxidized DPG. Since PPG includes primary and secondary alcohol groups, there might be two kinds of alcohol dehydrogenase included. In addition, an interesting question involves the broad substrate specificity of the bacterium toward (R,S)-PG structures included in atactic PPG and DPG. Enzymes involved in the metabolism of PPG have neither been purified nor characterized yet.

PTMG dehydrogenase was detected in the cell-free extracts of PTMG-utilizing bacteria, but it has not been characterized. As the PTMG-utilizing bacteria did not grow on PPG or PBG, secondary alcohols are possibly not metabolized. The PTMG-utilizing bacterium grew on PEG 400: when PTMG is oxidized to a carboxylic acid and then metabolized by β-oxidation, the resultant terminal structure $-O-CH_2COOH$ is similar to that obtained from PEG.

Anaerobic metabolism of PEG has been studied by Schink et al. [172–174]. They suggested that PEG acetaldehyde lyase might be involved in the metabolism of PEG, which is analogous to the diol dehydratase reaction, as shown in Fig. 9. The enzyme activity was not destroyed by proteinase K and found in the cytoplasm fraction. They suggested that at least one unmasked terminal hydroxyl group was necessary for the formation of the hemiacetal intermediate by transhydroxylation. The points to be confirmed are the further characterization of the enzyme and the permeability of a large molecule through the outer membrane and cell wall into the cytoplasm because their enzyme was not detected in the culture supernatant, the periplasm, or the membrane fraction. Dwyer and Tiedje also suggested that acetaldehyde was a direct metabolite of PEG [175]; they detected DCIP-dependent PEG dehydrogenase. Schink et al. thought that the level of PEG dehydrogenase activity was too low for a primary metabolic enzyme and it might work instead on impurities such as EG and acetaldehyde. These

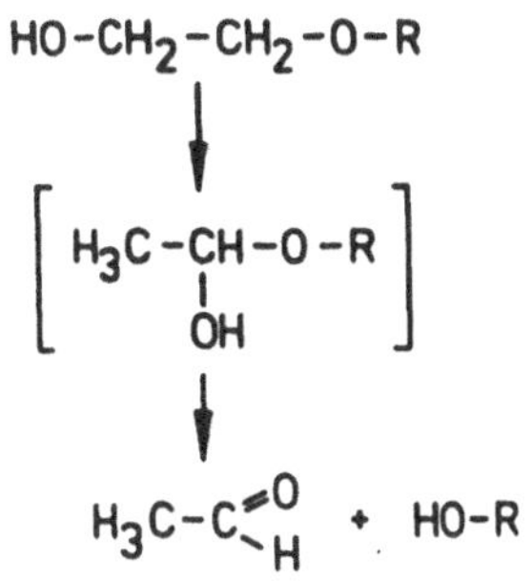

$$HO-CH_2-CH_2-O-R$$

$$\downarrow$$

$$\left[\begin{array}{c} H_3C-CH-O-R \\ | \\ OH \end{array}\right]$$

$$\downarrow$$

$$H_3C-C\begin{array}{c}{}^{\nearrow O}\\{}_{\searrow H}\end{array} \quad + \quad HO-R$$

Fig. 9. Anaerobic metabolism of PEG

results coincide with our results that degradation starts from the terminal of a long molecular chain and therefore at least one free alcohol group is necessary, although both aerobic and anaerobic mechanisms appear to be different and both ether-cleaving mechanisms necessitate further work.

4.3 Polyvinyl Alcohol

Two kinds of reactions are involved in the metabolism of PVA: oxidation and hydrolysis, as shown in Fig. 10. Extracellular inducible oxidases were purified from *Pseudomonas* O-3 [176, 177] and *Pseudomonas vesicularis* PD [178–180]. These enzymes were thought to be the only enzymes involved in the metabolism; it was assumed that 1,3-glycol structures in PVA were endogenously oxidized to diketones and then spontaneously hydrolyzed. Later the second enzyme was separated from the first enzyme and characterized as a diketone hydrolase. The oxidase was characterized as a secondary alcohol oxidase (SAO, EC 1.1.3.18), which included non-heme iron as prosthetic group [179, 180]. The enzyme required molecular oxygen and produced hydrogen peroxide. The enzyme oxidized hydroxyl groups of PVA at random and also linear aliphatic secondary alcohols to corresponding ketones. The activity was higher toward highly polymerized PVA as compared to the octamer [181].

A hydrolase was purified from *P. vesicularis* PD and named β-diketonehydrolase (BDH, EC 3.7.1.7) [182–184]. BDH hydrolyzed the C–C bond of oxidized PVA to produce a carboxylic acid and methyl ketone, as shown in Fig. 10. The enzyme also reacted on a linear aliphatic β-diketone such as 4,6-nonandione. Oxidized PVA was stable at room temperature under neutral conditions and required BDH for further degradation [185]. At greater than 50 °C and under acidic or alkaline conditions, oxidized PVA was unstable and spontaneously degraded to yield a carboxylic acid and methyl ketone similar to the enzymatic degradation. Pyrolysis was accelerated by phosphate. Huang et al. prepared biodegradable poly(enol-ketone) from PVA by oxidation [186].

The importance of membrane-bound PVA-degrading enzymes was suggested by Shimao et al. *Pseudomonas* sp. VM15C constitutively produced extracellular and intracellular oxidases [187]. Intracellular (membrane-bound) enzyme was

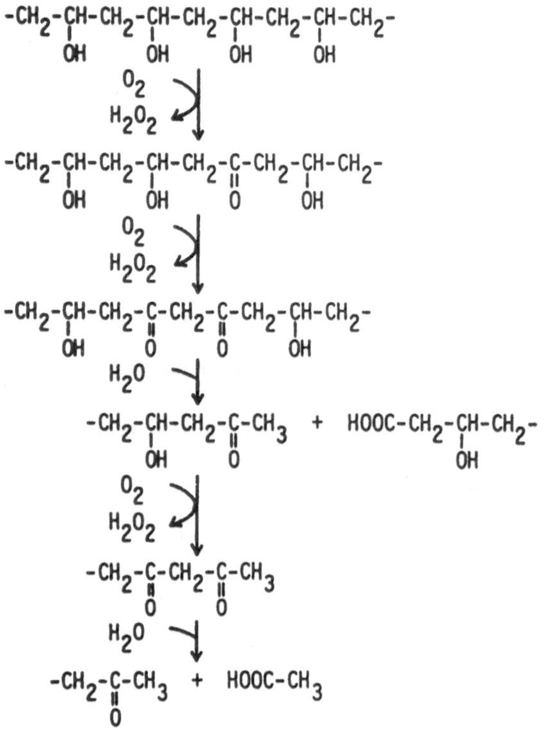

$$-CH_2-\underset{\underset{OH}{|}}{CH}-CH_2-\underset{\underset{OH}{|}}{CH}-CH_2-\underset{\underset{OH}{|}}{CH}-CH_2-\underset{\underset{OH}{|}}{CH}-CH_2-$$

O_2
H_2O_2

$$-CH_2-\underset{\underset{OH}{|}}{CH}-CH_2-\underset{\underset{OH}{|}}{CH}-CH_2-\underset{\underset{O}{\|}}{C}-CH_2-\underset{\underset{OH}{|}}{CH}-CH_2-$$

O_2
H_2O_2

$$-CH_2-\underset{\underset{OH}{|}}{CH}-CH_2-\underset{\underset{O}{\|}}{C}-CH_2-\underset{\underset{O}{\|}}{C}-CH_2-\underset{\underset{OH}{|}}{CH}-CH_2-$$

H_2O

$$-CH_2-\underset{\underset{OH}{|}}{CH}-CH_2-\underset{\underset{O}{\|}}{C}-CH_3 \quad + \quad HOOC-CH_2-\underset{\underset{OH}{|}}{CH}-CH_2-$$

O_2
H_2O_2

$$-CH_2-\underset{\underset{O}{\|}}{C}-CH_2-\underset{\underset{O}{\|}}{C}-CH_3$$

H_2O

$$-CH_2-\underset{\underset{O}{\|}}{C}-CH_3 \quad + \quad HOOC-CH_3$$

Fig. 10. Aerobic metabolic pathway of PVA

characterized as a secondary alcohol oxidase. PVA dehydrogenase was membrane-bound and required PQQ as a prosthetic group [188]. The enzyme was purified and characterized as a secondary alcohol dehydrogenase. They suggested that the membrane-bound dehydrogenase was linked with a respiratory chain.

4.4 Polyacrylate

The metabolic pathway for PA was proposed on the basis of acyl-coenzyme A (Co A) synthetase activities and metabolic products from PTCA [55, 56]. Fatty acids and acrylic acid are known to be activated by CoA and then metabolized by β-oxidation. Acyl-CoA synthetase activity towards PTCA, PA 500 and PA 1000 was detected with dialyzed cell-free extracts of PTCA-utilizing bacteria, as shown in Table 10. These results suggested that PTCA-utilizing bacteria could metabolize PTCA and PA by the same process. To characterize the metabolic products, culture supernatants of PTCA-utilizing bacteria were analyzed by HPLC. As bacteria grew on PTCA, the pH of the culture fluid became alkaline suggesting consumption of the acid. The main products formed were in fraction A and a small amount of fraction B. Although a large amount of fraction A was

Table 10. Acyl-CoA synthetase activities towards oligomeric and polymeric acrylates

Strain	Specific Activity (U/mg)[a] Substrate (25 mM)[b]		
	PTCA	PA 500	PA 1000
Microbacterium sp. II–7–12	0.015	0.032	0.050
Xanthomonas maltophilia WI	0.014	0.047	0.070
Acinetobacter genosp. 11 W2	0.039	0.063	0.133

[a] Acyl-CoA synthetase activity was measured with dialyzed cell-free extracts and expressed as units per milligram protein.
[b] The substrate concentration (25 mM) corresponded to approximately 0.5% PTCA, 1.25% PA 500 and 2.5% PA 1000.
Reprinted from Ref. [55].

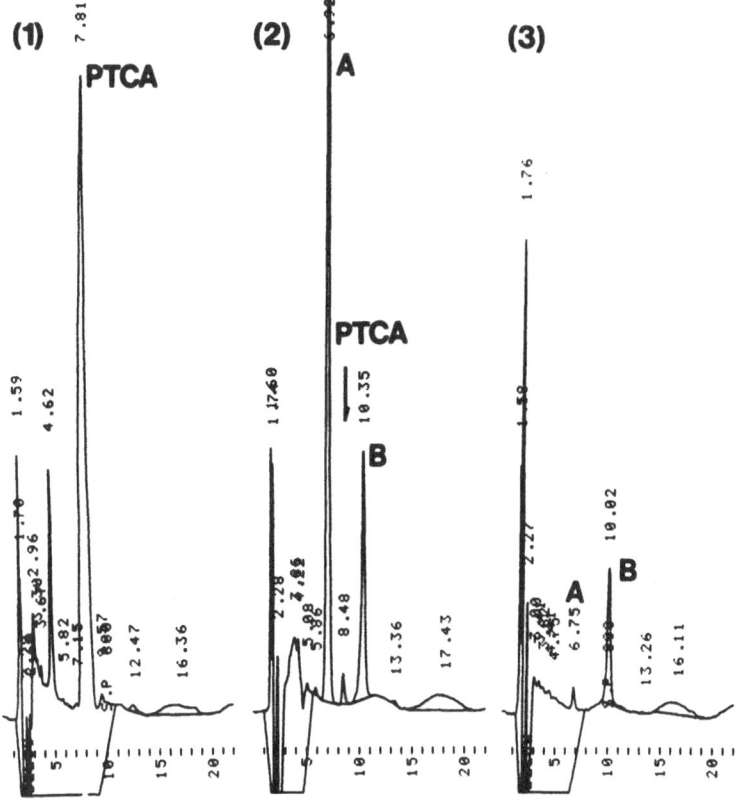

Fig. 11. Time course for the consumption of PTCA and the amounts of fractions A and B formed by growing cells of *Microbacterium* sp. (1) Autoclaved medium; (2) 3-day culture supernatant; (3) 7-day culture supernatant. Reprinted from Ref. [56]

formed in 2–3 days, it was quickly metabolized further and disappeared from the culture supernatant, as shown in Fig. 11. Fraction B was only slowly metabolized and remained in the culture supernatant for an extended time (5–7 days). These metabolites were analyzed on a Waters $5C_8$ column by LC-MS. From mass spectra, the molecular size of fractions A and B was calculated to be 202 which corresponds to the molecular weight minus two of the substrate, which is 204. These compounds were consistent with 1,3,5-(1 or 2-pentene)tricarboxylic acid. Another peak (C) was detected on Cosmosil 5CN-R column and characterized by LC-MS to be 1,3,5-(2-oxopentane)tricarboxylic acid. These results suggested that enzymes similar to those for β-oxidation are involved in the metabolism of PTCA, as shown in Fig. 12. Considering that a β-oxidation mechanism is working for PTCA, fraction A might correspond to 1,3,5-(1-pentene)tricarboxylic acid and fraction B to 1,3,5-(2-pentene)tricarboxylic acid. There still remain two possibilities for proceeding with the metabolic pathway of PTCA: (1) decarboxylation of a depolymerized compound by one acrylic unit followed by an ordinary β-oxidation process to liberate acetic acid and (2) repeated liberation of malonic acid from the depolymerized compound. Malonic acid is known as a potent inhibitor of succinate dehydrogenase, a key enzyme in the Krebs cycle, in eukaryotic cells. As three kinds of PTCA-utilizing bacteria utilized

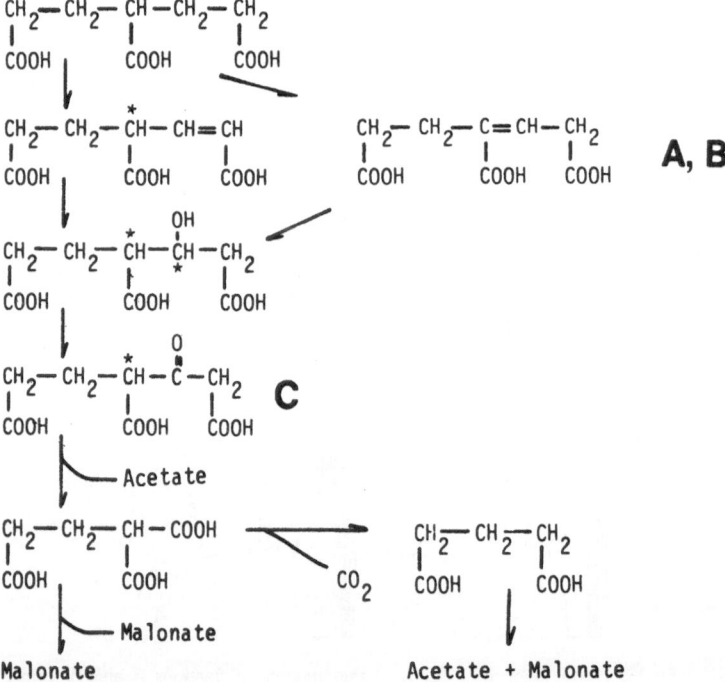

Fig. 12. The proposed metabolic pathway for PTCA. Reprinted from Ref. [56]

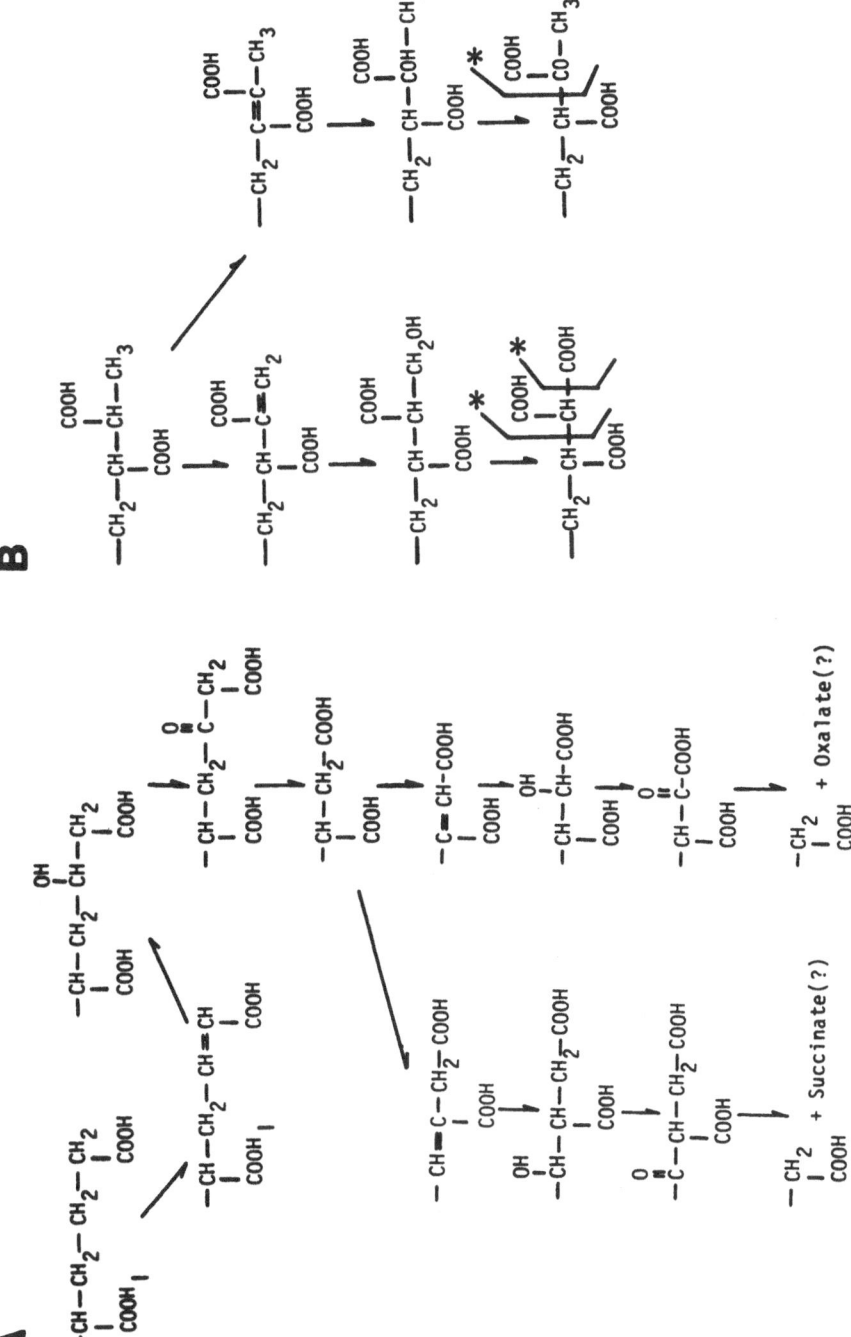

Fig. 13. The proposed metabolic pathway for a head-to-head structure (A) and a tail-to-tail structure (B) of PA. *Possible cleaving points. Reprinted from Ref. [56]

malonic acid as a growth substrate, malonic acid is nontoxic to these prokary-
otic cells. The metabolism of PA seems to be similar to that of PTCA. The
metabolic pathway for head-to-head and tail-to-tail structures of a polymer was
inferred by analogy with the metabolism for a head-to-tail structure found in
PTCA, as shown in Fig. 13. As fraction B was very slowly metabolized, we
cannot deny that a double bond is formed at positions other than the α- and
β-positions. In other words, when considering a polymer, a short length of the
chain might be penetratable through the outer membrane or cell wall, although
a long chain with polycarboxylic acids could hardly reach the cytoplasmic membrane. PA 1000–3000 was metabolized by washed cells. As the oxidation site is
known to be located on the bacterial membrane, a polymer has to come into
contact with metabolizing enzymes on a cytoplasmic membrane. Actually the
degradation of PTCA by washed cells was completely inhibited by 5 mM NaN_3,
an inhibitor of the respiratory chain, suggesting that the metabolism of PTCA
proceeds on the cytoplasmic membrane. Thus, it is quite reasonable to think
that the degradation mode of PA is substantially exogenous, and not randomly
endogenous.

From the comparison with the metabolism of PA, polymethacrylate might not
be metabolized, because a methyl group might inhibit the formation of a double
bond. Hayashi et al. reported that their acrylic oligomer-utilizing bacterium did
not degrade methacrylic oligomers [57].

5 Generalizations Concerning Biodegradation

From the data described above, we can make some general guidelines for the
relationship between structure and biodegradation as summarized by Swift [19]

1) Naturally occurring polymers are biodegradable. Chemically modified natural
   polymers may biodegrade, depending on the extent of modification and the
   kind of modifying group.
2) Synthetic addition polymers with carbon-chain backbones do not biodegrade at
   molecular weights greater than about 1000. Polyvinyl alcohol is an exception,
   the biodegradability of which is due to pendant hydroxyl groups which are
   readily converted to hydrolyzable carbonyl groups.
3) Synthetic addition polymers with hetero-atoms in their backbones may biodegrade; these include polyacetals and polyesters.
4) Synthetic step-growth or condensation polymers are generally biodegradable
   to a greater or lesser extent, depending on:
   chain coupling (ester > ether > amide > urethane);
   molecular weight (lower is faster than higher);
   morphology (Tm) (amorphous is faster than crystalline);

hardness (Tg) (softer is faster than harder)
and
hydrophilicity vs hydrophobicity (hydrophilic is faster than hydrophobic).

5) Water solubility does not guarantee biodegradability.

Thus, biodegradability is primarily dependent on hydrolyzable and oxidizable chemical structures, balance of hydrophobicity, and molecular weights. Optical selectivity is either strict or loose, depending on the microorganism. Physical properties such as crystallinity, orientation, Tm or Tg, and morphological properties such as surface area or thickness affect the rate of degradation.

Recalcitrant polymers may be rendered susceptible to degradation by

a) incorporation into carbon backbones of weak linkages which are hydrolytically unstable, e.g. ester, photoreactive ketones;
b) blending with biodegradable additives such as starch;
c) grafting or copolymerizing with biodegradable segments;
d) addition of activators to promote oxidative degradation;
and
e) keeping the recalcitrant part as short as possible.

It should be borne in mind that blending, grafting or copolymerizing means only the collapse of the morphology of the polymer and the removal of a biodegradable part, instead of total removal of the whole polymer.

Important factors relating to enzymes which are involved in the metabolism of synthetic polymers are as follows:

(a) extracellular or endocellular location (extracellular is faster than endocellular).
(b) single enzyme or multi-enzymes (single is faster than multi).
(c) substrate specificities (broader is better than narrower).
(d) affinity for substrates (higher is better than lower).
(e) endogenous or exogenous actions (endogenous is faster than exogenous).
(f) specific or ubiquitous (ubiquitous is better than specific).

When degradation is catalyzed by a single enzyme, an extracellular enzyme of endogenous nature is best; e.g., esterases. Endocellular enzymes cannot act on long molecules unless they are located in the periplasmic space or membrane-bound and a terminal group of the long molecule is transported into the periplasmic space or onto the surface of the cytoplasmic membrane. When metabolizing enzymes are linked with a respiratory chain, they have to be located in the periplasmic space or be membrane-bound. Thus, exogenous reactions seem to be advantageous for biodegradation of a long chain structure by endocellular enzymes, because only a terminal group of the long molecule has to get through the cell wall or outer membrane to contact the enzymes, instead of transport of the whole molecule; e.g., PEG of high molecular weight. The metabolizing enzymes so far reported seem to have extended their specificity not only towards naturally occurring polymers, but also toward synthetic polymers having structures similar to natural compounds. Therefore, these enzymes may have evolved from the original enzymes which can metabolize only natural materials. This assumption is supported by results indicating that these enzymes act on analogous materials smaller than polymers and oligomers.

Finally I would like to emphasize that the importance of mixed cultures will increase in this area; recalcitrant xenobiotic compounds are often degraded by the concerted action of many bacteria. Biodegradable polymers made by blending, copolymerizing or grafting include many chemical structures whose degradations necessitate many enzyme reactions which have to be performed by a community of microorganisms.

6 References

1. Gu JD, McCarthy SP, Smith GP, Eberiel D, Gross RJ (1992) Polym Mater Sci Eng 67: 351
2. Buchanan CM, Gardner RM, Komarek RJ, Gedon SC, White AW (1993) Biodegrad Polym Packag: 133; CA 119: 187832q
3. Yano S, Hirose S, Hatakeyama H (1989) The mechanical properties of polyurethane from lignocellulose. In: Kennedy JF, et al. (eds) Wood Processing and Utilization. Ellis Horwood, Chichester; p 263
4. Yasui M, et al. (1991) Kobunshi Kako (in Japanese) 40: 407
5. Jane J, Lim S-T, Paetau I (1993) Degradable plastics made from starch and protein. In: Ching C, Kaplan D, Thomas E (eds) Biodegradable Materials and Packing Proceeding, Technomic, Lancaster, p. 63
6. Stevens RA, Levin RE (1976) Appl Environ Microbiol 31: 896
7. Linker A, Evans LR (1984) J Bacteriol 159: 958
8. Lemoigne M (1925) Ann Inst Pasteur (Paris) 39: 144
9. Imperial Chemical Industries Eur Pat 0052459 (1981); 0069497 (1983)
10. Byrom D (1987) Trends Biotechnol 5: 246
11. Doi Y et al. (1987) J Chem Soc Chem Commun 9: 1635
12. Kunioka M, et al. (1988) Polymer Commun 29: 174
13. Kunioka M, et al. (1989) Appl Microbiol Biotechnol 30: 569
14. Hiramatsu M, Doi Y (1993) Polymer 34: 4782
15. Brandl H, Gross RA, Lenz RW, Fuller RC (1988) Appl Environ Microbiol 54: 1977
16. Ramsay BA, et al. (1993) Appl Environ Microbiol 59: 1242
17. Poirier Y, et al. (1992) Science 256: 520
18. Byrom D (1991) Biomaterials, Macmillan, New York
19. Swift G (1993) Accnts Chem Res 26: 105
20. Berkeley RCW, Gooday GW (1979) Microbial Polysaccharides and Polysaccharases, Academic Press, New York
21. Ivanovics G, Erdos LZ (1937) Z Immunitatsforsch 90: 5&304
22. Bovarnick M (1942) J Biol Chem 145: 15
23. Fujii H (1963) J Agr Chem Soc Japan (in Japanese) 37: 407
24. Hara T, et al. (1982) J Appl Biochem 4: 112
25. Goto A, Kunioka M (1992) Biosci Biotech Biochem 56: 1031
26. Kream J, et al. (1954) Arch Biochem Biophys 53: 334
27. Volcami BB, et al. (1957) J Bacteriol, 74: 646
28. Troy, FA (1973) J Biol Chem 248: 316
29. Shima S, Sakai H (1977) Agr Biol Chem 41: 1807
30. Shima S, et al. (1984) J Antibiot 37: 1449
31. Simada K, Matsushima K (1967) J Agr Chem Soc Japan (in Japanese) 41: 454
32. Nagata N, Nakahara T, Tabuchi T (1993) Biosci Biotech Biochem 57: 638

33. Vert M, Lenz RW (1978) Polym Prepr 20: 608
34. Abe Y, Matsumura S, Imai K (1986) Yukagaku (in Japanese) 35: 937
35. Ohtani N, Kimura R, Kitao T (1987) Kobunshi Ronbunshu 44: 701
36. (1990) Bioproc Technol 12(11): 7
37. Ishizaki A, et al. (1992) Biotechnol Lett, 14: 599
38. (1988) Chem Week Sept 7: 32
39. Rypacek F, Saudek V, Pytela J, Skarda V, Drobnik J (1985) Macromol Chem Suppl 9: 129
40. Donlar Corp (1992) Bioproc Technol 14: 1
41. Alford DD, Wheeler AP, Pettigrew CA (1994) J Environ Polym Degr 2: 225
42. Huang SJ, Roby MS, Macri CA, Cameron JA (1992) The Effects of Structure and Morphology
    on the Degradation of Polymers with Multiple Groups. In: Vert M, et al. (ed) Biodegradable
    Polymers and Plastics, Royal Society of Chemistry, London, p. 149
43. Albertsson AC, Ranby B (1976) In: Sharpley JM, Kaplan AM (ed) Proc Third Int Biodegr
    Symp, Applied Science, London, p. 743
44. Potts JE, Clendinning, Ackart WB, Niegisch WD (1973) In: Guillet J (ed) Polymers and Eco-
    logical Problems, Plenum Press, New York, p. 61
45. Albertsson AC, Andersson SO, Karlsson S (1987) Polym Degr Stability 18: 73
46. Albertsson AC (1993) J.M.S. Pure Appl Chem A30 (9&10): 757
47. Scott G (1975) Polym Age 6: 54
48. Guillet J, Huber HX, Scott J (1993) Abstr 2nd Nat Mtg Bio/Environ Degrad Polym Soc, Chicago
    p. 26
49. (1993) Chem Week, Apr 21: 15
50. Tsuchii A, Suzuki T, Takahara Y (1977) Agr Biol Chem 41: 2417
51. Tsuchii A, Suzuki T, Takahara Y (1978) Agr Biol Chem 42: 1217
52. Hosoya H, Miyazaki N, Sugisaki Y, Takanashi E, Tsurufuji M, Yamasaki M, Tamura G (1978)
    Agr Biol Chem 42: 1545
53. Tsuchii A, Suzuki T, Takahara Y (1979) Agr Biol Chem 43: 2441
54. Matsumura S, Maeda S, Takahashi J, Yoshikawa S (1988) Kobunshi Ronbunshu 45: 317
55. Kawai F (1993) Appl Microbiol Biotechnol 39: 382
56. Kawai F (1994) J Environ Polym Degr 2: 59
57. Hayashi T, Mukouyama M, Sakano K, Tani Y (1993) Appl Environ Microbiol 59: 1555
58. Hayashi T, Nishimura H, Sakano K, Tani Y (1994) Biosci Biotech Biochem 58: 444
59. Rittmann BE, Sutfin JA, Henry B (1992) Biodegr 2: 181
60. Suzuki T, Ichihara Y, Yamada M, Tonomura K (1973) Agric Biol Chem 37: 747
61. Suzuki T, Ichihara Y, Dazai M, Misono M (1973) Hakko Kogaku Kaishi 51: 692
62. Watanabe Y, Morita M, Hamada N, Tsujisaka Y (1975) Agric Biol Chem 39: 2447
63. Sakai K, Hamada N, Watanabe Y (1987) Kagaku To Kogyo (Osaka) 61: 372
64. Sakazawa C, Shimao M, Taniguchi Y, Kato N (1981) Appl Environ Microbiol 41: 261
65. Shimao M, Yamamoto H, Ninomiya K, Kato N, Adachi O, Ameyama M, Sakazawa C (1984)
    Agric Biol Chem 48: 2873
66. Hashimoto S, Fujita M (1985) J Ferment Technol 63: 471
67. Matsumura S, Kurita H, Shimokobe H (1993) Biotech Lett 15: 749
68. Darby RT, Kaplan AM (1968) Appl Microbiol 16: 900
69. Potts JE, Clendinning RA, Ackart WB, Niegisch (1972) Am Chem Soc, Polymer Preprints
    13: 629
70. Fields RD, Rodriguez F, Finn RK (1974) J Appl Polym Sci 18: 3571
71. Huang SJ, et al. (1976) Proc 3rd Int Biodegradation Symp, p. 731
72. Tokiwa Y, Suzuki T (1974) J Ferment Technol 52: 393
73. Tokiwa Y, Suzuki T (1976) J Ferment Technol 54: 603
74. Tokiwa Y, Suzuki T (1977) Agr Biol Chem 41: 265
75. Tokiwa Y, Suzuki T (1977) Nature 270: 76
76. Tokiwa Y, Suzuki T (1988) Agr Biol Chem 52: 1937
77. Tokiwa Y, et al. (1990) Biodegradation of Synthetic Polymers Containing Ester Bonds. In:
    Glass JE, Swift G (ed) Agricultural and Synthetic Polymers, Am Chem Soc, Washington, DC,
    p. 136
78. Takiyama E, Fujimaki T (1993) Abstr 3rd Int Sci Workshop Biodegrad Plastics and Polymer,
    Osaka, p. 34
79. Yamagishi K, et al. (1973) Chem Lett (Tokyo): 692
80. Monsanto, U.S. Patents 4,144,226; 4,146,495; 4,204,652; 4,233,422; 4,233,423

81. Kawai F (1987) CRC Crit Rev Biotechnol 6: 273
82. Fincher EL, Payne WJ (1962) Appl Microbiol 10: 542
83. Takeuchi M, Kawai F, Shimada Y, Yokota A (1993) System Appl Microbiol 16: 227
84. Kawai F, Yamanaka H (1986) Arch Microbiol 146: 125
85. Mills EJ, Stack VT (1954) Eng Bull Pardue Univ Eng Ext Ser 87: 449
86. Shink B, Stieb M (1983) Appl Environ Microbiol 45: 1905
87. Dwyer DF, Tiedje JM (1983) Appl Environ Microbiol 46: 185
88. Grant MA, Payne WJ (1983) Biotech Bioeng 25: 627
89. Taylor BF, Campbell WL, Chinoy I (1970) J Bacteriol 102: 430
90. Watson GK, Jones N (1977) Water Res 11: 95
91. Kawai F, Hanada K, Tani Y, Ogata K (1977) J Ferment Technol 55: 89
92. Kawai F (1993) Kobunshi Ronbunshu 50: 775
93. Kawai F, Okamoto T, Suzuki T (1985) J Ferment Technol 63: 239
94. Kawai F, Moriya F (1991) J Ferment Bioeng 71: 1
95. Ebata M, Morita K (1959) J Biochem 46: 407
96. Iizuka H, Tanabe I, Fukumura T, Kato K (1967) J Gen Appl Microbiol 13: 125
97. Nonomura S, Kotsani R, Urakabe R, Shima S, Sakai H (1974) Agr Biol Chem 38: 1755
98. Kinoshita S, Kageyama S, Iba K, Yamada Y, Okada H (1975) Agr Biol Chem 39: 1219 (1975)
99. Kinoshita S, Negoro S, Muramatsu M, Bisaria VS, Sawada S, Okada H (1977) Eur J Biochem 80: 489
100. Negoro S, et al. (1980) J Bacteriol 143: 238
101. Negoro S, Taniguchi T, Kanaoka M, Kimura H, Okada H (1983) J Bacteriol 155: 22
102. Okada H, Negoro S, Kimura H, Nakamura S (1983) Nature 306: 203
103. Shuttleworth WA, Seal KJ (1986) Appl Microbiol Biotechnol 23: 407
104. Seal KJ, Pathirana RA (1982) Internat Biodeter Bull 18: 81
105. Steinmann HW (1970) Am Chem Soc Polym Prepr 11: 285
106. Lynn MM, Stannett VT, Gilbert RD (1980) J Polym Sci Polym Chem Ed, 18: 1976
107. Marhaut RE, Zhao Q, Anderson JM, Hiltner A (1987) Polymer 28: 2032&2040
108. Matsumura S, Yoshikawa S (1990) Biodegradable Poly(carboxylic acid) Design. In: Glass JE, Swift G (ed) Am Chem Soc Symp Ser 433, Am Chem Soc, Washington, DC, p. 124
109. Nakayama A, et al. (1990) Abstr Int Symp Biodegrad Polym, Tokyo, p. 171
110. Yasuda T, Aida T, Inoue S (1984) Macromolecules 17: 2217
111. Kawai F (1992) Mechanisms of bacterial degradation of polyethers and their copolymers. In: Vert M, et al. (ed) Biodegradable Polymers and Plastics, The Royal Soc Chem, Cambridge, p. 20
112. Avella M, Martuscelli E (1988) Polymer 29: 1731
113. Kumagai Y, Doi Y (1993) J Environ Polym Degr 1: 81
114. Cohn D (1989) Abstr 197th ACS Nat Mtg, BTEC No. 47
115. Albertsson AC, Ljungquist O (1986) J Macromol Sci A25: 467
116. Bastioli C, Bellotti V, Del Gindice L, Gilli G (1993) J Environ Polym Degr 1: 181
117. Bailey WJ, et al. (1986) Makromol Chem Makromol Symp 6: 81
118. Otey FH, Westhoff RP, Russell CR (1977) Ind Eng Chem Prod Res Dev 16: 305
119. Otey FH, Westhoff RP, Doane WM (1980) Ind Eng Chem Prod Res Dev 19: 592
120. Otey FH, Westhoff RP, Doane WM (1987) Proc SPI Symp Degr Plastics, Soc Plastics Ind, Washington, DC, p. 39
121. Griffin GJL (1972) US Patent 4,021,388
122. Griffin GJL (1975) Am Chem Soc Adv Chem Ser 134
123. Griffin GJL (1987) Proc SPI Symp Degr Plastics, Soc Plastics Ind, Washington, DC, p. 47
124. Wool RP, Cole MA (1988) Am Soc Microbiol Eng Handbook 2: 783
125. Guillet J, Scott G (1973) In: Guillet J (ed) Polymers and Ecological Problems. Plenum Press, New York, 1: 27
126. U.S.D.A., U.S. Patent 3,949,145
127. Coloro'll, British Patent 824821
128. Dennenburg RJ, et al. (1978) J Appl Polym Sci Chem A22: 459
129. Potts JE, U.S. Patents 3,901,838; 3,921,333
130. Chisso Corp, Eur Patent 252555-A
131. Techocolor Celebran, French Patent 2611732-A
132. Griffin GJL, World Patent 8,809,354-A
133. National Technology Information Service, PB 89-857114; 88-863295

134. Potts JE, et al. (1978) In: Aspects of Degr Stabili Polym, Elsevier, Amsterdam, p. 617
135. Kendirick JP, Diss Abst 82329391
136. Brown AJ, Kolattukudy PE (1978) Arch Biochem Biophys 190: 17
137. Chowdhury AA (1963) Arch Microbiol 47: 167
138. Muller B, Jendrossek D (1993) Appl Microbiol Biotechnol 38: 487
139. Tanio T, Fukui F, Shirakura Y, Saito T, Tomita K, Kaiho T, Masamune S (1982) Eur J Biochem 127: 71
140. Shirakura Y, Fukui T, Saito T, Okamoto Y, Narikawa T, Koide K, Tomita K, Takemasa Y, Masamune S (1986) Biochim Biophys Acta 880: 46
141. Jendrossek D, Knoke I, Habibian RB, Steinbuchel A, Schlegel HG (1993) J Environ Polymer Degrad 1: 53
142. Yamada K, Mukai K, Doi Y (1993) Int J Biol Macromol 15: 215
143. Mukai K, Yamada K, Doi Y (1993) Int J Biol Macromol 15: 361
144. Saito T, Shiraki M, Shimada T, Tatsumichi M (1993) Abstr 3rd Int Sci Workshop Biodegrad Plastics and Polymers, Osaka, p. 21
145. Brucato CL, Wong SS (1991) Arch Biochem Biophys 290: 497
146. Schirmer A (1993) Appl Environ Microbiol 59: 1220
147. Fukui T, Narikawa T, Miwa K, Shirakura Y, Saito T, Tomita K (1988) Biochim Biophys Acta 952: 164
148. Mukai K, Doi Y, Sema Y, Tomita K (1993) Int J Biol Macromol 15: 601
149. Saito T, Suzuki K, Yamamoto J, Fukui T, Miwa K, Tomita K, Nakanishi S, Odani S, Suzuki J, Ishikawa K (1989) J Bacteriol 171: 184
150. Jendrossek D, Muller B, Schlegel HG (1993) Eur J Biochem 218: 701
151. Kurusu Y, Kohama K, Uchida Y, Saito T, Yukawa H (1993) Abstr 3rd Int Sci Workshop on Biodegrad Plastics and Polymers. Osaka, p. 58
152. Janssen PH, Harfoot CG (1990) Arch Microbiol 154: 253
153. Budwill K, Fedorak PM, Page WJ (1992) Appl Environ Microbiol 58: 1398
154. Harada T, Nagashima Y (1975) J Ferment Technol 53: 218
155. Patterson SJ, Scott CC, Tucker KBE (1970) J Am Oil Chem Soc 47: 37
156. Herold DA, Rodeheaver GT, Bellamy WT, Fitton LA, Bruns DE, Edlich RF (1982) Toxicol Appl Pharmacol 65: 329
157. Hosoya H, Miyazaki N, Sugisaki Y, Takanashi E, Tsurufuji M, Yamasaki M, Tamura G (1978) Agric Biol Chem 42: 1545
158. Kawai F, Kimura T, Fukaya M, Tani Y, Ogata K, Ueno T, Fukami H (1978) Appl Environ Microbiol 35: 679
159. Kawai F, Kimura T, Tani Y, Yamada H, Ueno T, Fukami H (1983) Agric Biol Chem 47: 1669
160. Haines JR, Alexander M (1975) Appl Microbiol 29: 621
161. Pearce BA, Heydeman MT (1980) J Gen Microbiol 118: 21
162. Schöberl P (1983) Tenside Deterg 20: 57
163. Schöberl P (1985) Tenside Deterg 22: 70
164. Thélu J, Medina L, Pelmont J (1980) FEMS Microbiol Lett 8: 187
165. Obradors N, Aguilar J (1991) Appl Environ Microbiol 57: 2383
166. Kawai F, Yamanaka H (1989) Ferment Bioeng 67: 300
167. Kawai F, Kimura T, Tani Y, Tamada H, Kurachi M (1980) Appl Environ Microbiol 40: 701
168. Yamanaka H, Kawai F (1989) J Ferment Bioeng 67: 324
169. Kawai F, Yamanaka H, Ameyama M, Shinagawa E, Matsushita K, Adachi O (1985) Agric Biol Chem 49: 1071
170. Yamanaka K (1991) Agric Biol Chem 55: 837
171. Kawai F (1985) FEMS Microbiol Lett 30: 273
172. Wagener S, Schink B (1988) Appl Environ Microbiol 54: 561
173. Schramm E, Schink B (1991) Biodegr 2: 71
174. Frings J, Schramm E, Schink B (1992) Appl Environ Microbiol 58: 2164
175. Dwyer DF, Tiedje JM (1986) Appl Environ Microbiol 52: 852
176. Suzuki T (1976) Agric Biol Chem 40: 497
177. Suzuki T (1978) Agric Biol Chem 42: 1187
178. Watanabe Y, Hamada N, Morita M, Tsujisaka Y (1976) Arch Biochem Biophys 174: 575
179. Morita M, Hamada N, Sakai K, Watanabe Y (1979) Agric Biol Chem 43: 1225
180. Sakai K, Hamada N, Watanabe Y (1985) Agric Biol Chem 49: 817
181. Sakai K, Hamada N, Watanabe Y (1985) Agric Biol Chem 50: 989

182. Sakai K, Hamada N, Watanabe Y (1985) Agric Biol Chem 49: 1901
183. Sakai K, Morita M, Hamada N, Watanabe Y (1981) Agric Biol Chem 45: 63
184. Sakai K, Hamada N, Watanabe Y (1985) Agric Biol Chem 49: 827
185. Sakai K, Hamada N, Watanabe Y (1984) Agric Biol Chem 48: 1093
186. Huang SJ, Wang IF, Quinga E (1983) Polym Sci Technol 21: 75
187. Shimao M, Tsuda T, Takahashi M, Kato N, Sakazawa C (1983) FEMS Microbiol Lett 20: 429
188. Shimao M, Ninomiya K, Kano O, Kato N, Sakazawa C (1986) Appl Environ Microbiol 51: 268
189. Gould JM, Gordon SH, Dexter LB, Swanson CL (1990) Biodegradation of Starch-Containing Plastics. In: Glass JE, Swift G (ed) Am Chem Soc Symp Ser 433, Am Chem Soc, Washington, DC, p. 65
190. Kawai F (1994) Biodegradation of polyethers and polyacrylate. In: Doi Y, Fukuda K (eds) Biodegrad Plastics and Polymers. Elsevier, Amsterdam, p. 24

Author Index Volume 52

Author Index Vols. 1-50 see Vol. 50

Subject Index

Springer-Verlag
and the Environment

We at Springer-Verlag firmly believe that an international science publisher has a special obligation to the environment, and our corporate policies consistently reflect this conviction.

We also expect our business partners – paper mills, printers, packaging manufacturers, etc. – to commit themselves to using environmentally friendly materials and production processes.

The paper in this book is made from low- or no-chlorine pulp and is acid free, in conformance with international standards for paper permanency.